Recommended Research Priorities for the Qatar Foundation's Environment and Energy Research Institute

Nidhi Kalra, Obaid Younossi, Kristy N. Kamarck, Sarah Al-Dorani, Gary Cecchine, Aimee E. Curtright, Chaoling Feng, Aviva Litovitz, David R. Johnson, Mohammed Makki, Shanthi Nataraj, David S. Ortiz, Parisa Roshan, Constantine Samaras

Sponsored by the Qatar Foundation for Education, Science, and Community Development

 RAND-QATAR POLICY INSTITUTE

This research was sponsored by the Qatar Foundation for Education, Science, and Community Development and was conducted in the RAND-Qatar Policy Institute and the Environment, Energy, and Economic Development Program within RAND Infrastructure, Safety, and Environment.

Library of Congress Cataloging-in-Publication Data is available for this publication

ISBN: 978-0-8330-5820-1

The RAND Corporation is a nonprofit institution that helps improve policy and decisionmaking through research and analysis. RAND's publications do not necessarily reflect the opinions of its research clients and sponsors.

RAND® is a registered trademark.

Published 2011 by the RAND Corporation
1776 Main Street, P.O. Box 2138, Santa Monica, CA 90407-2138
1200 South Hayes Street, Arlington, VA 22202-5050
4570 Fifth Avenue, Suite 600, Pittsburgh, PA 15213-2665
RAND URL: http://www.rand.org/
To order RAND documents or to obtain additional information, contact
Distribution Services: Telephone: (310) 451-7002;
Fax: (310) 451-6915; Email: order@rand.org

Principal Investigators
Obaid Younossi and Nidhi Kalra

Phase 1: Survey of Gulf Cooperation Council	Phase 2: Research Priorities
Kristy N. Kamarck (lead)	Nidhi Kalra (lead)
Sarah Al-Dorani	Gary Cecchine
Mohammed Makki	Chaoling Feng
Parisa Roshan	Constantine Samaras
Obaid Younossi	Obaid Younossi

Research Topic Authors

Natural gas production and processing	David S. Ortiz
Petroleum production and processing	David S. Ortiz
Carbon capture and storage	Aimee E. Curtright
Solar energy development	Aimee E. Curtright
Fuel cells	Aimee E. Curtright
Green buildings	Aviva Litovitz and Constantine Samaras
Smart grids	Aviva Litovitz and Aimee E. Curtright
Strategic energy planning	Shanthi Nataraj
Desalination	Shanthi Nataraj
Groundwater sustainability	Gary Cecchine and Shanthi Nataraj
Water demand management	Shanthi Nataraj
Integrated water resource management	Shanthi Nataraj
Environmental characterization	Shanthi Nataraj and David R. Johnson

Preface

Qatar is developing rapidly. As a consequence, its population is also growing rapidly. Industrial development and population growth have increased Qatar's energy, water, and other resource needs substantially and threaten its environmental sustainability. In response, Qatar's leadership has created a vision of sustainable development for the country, and the Qatar Foundation for Education, Science, and Community Development (QF) is establishing a national research institute that conducts and collaborates on applied research in energy, environment, and water issues to help achieve this vision. QF asked the RAND-Qatar Policy Institute (RQPI) to recommend research priorities based on Qatar's needs and goals in these areas and based on collaboration opportunities in the Cooperation Council for the Arab States of the Gulf (also known as the Gulf Cooperation Council, or GCC) and beyond. This monograph responds to that request. The findings were originally briefed to QF in November 2010. The intended audience includes officials in QF and the Qatari government. This work might also be of interest to researchers and administrators at other research institutions in Qatar, the GCC region, and abroad.

The project that produced this monograph was conducted under the auspices of RQPI and in the Environment, Energy, and Economic Development Program (EEED) within RAND Infrastructure, Safety, and Environment (ISE). RQPI is a partnership of the RAND Corporation and QF. The aim of RQPI is to offer the RAND style of rigorous and objective analysis to clients throughout the greater Middle East. In serving clients in the Middle East, RQPI draws on the full professional resources of the RAND Corporation. Related RAND research has yielded the following publications:

- Victoria A. Greenfield, Debra Knopman, Eric Talley, Gabrielle Bloom, Edward Balkovich, D. J. Peterson, James T. Bartis, Stephen Rattien, Richard A. Rettig, Mark Y. D. Wang, Michael Mattock, Jihane Najjar, and Martin C. Libicki, *Design of the Qatar National Research Fund: An Overview of the Study Approach and Key Recommendations*, Santa Monica, Calif.: RAND Corporation, TR-209-QF, 2008
- David S. Ortiz, Aimee E. Curtright, Constantine Samaras, Aviva Litovitz, and Nicholas Burger, *Near-Term Opportunities for Integrating Biomass into the U.S. Electricity Supply: Technical Considerations*, Santa Monica, Calif.: RAND Corporation, TR-984-NETL, 2011
- Michael Toman, James Griffin, and Robert J. Lempert, *Impacts on U.S. Energy Expenditures and Greenhouse-Gas Emissions of Increasing Renewable-Energy Use*, Santa Monica, Calif.: RAND Corporation, TR-384-1-EFC, 2008
- Tom LaTourrette, Thomas Light, Debra Knopman, and James T. Bartis, *Managing Spent Nuclear Fuel: Strategy Alternatives and Policy Implications*, Santa Monica, Calif.: RAND Corporation, MG-970-RC, 2010.

The RAND Environment, Energy, and Economic Development Program

This research was conducted within EEED. The mission of RAND Infrastructure, Safety, and Environment is to improve the development, operation, use, and protection of society's essential physical assets and natural resources and to enhance the related social assets of safety and security of individuals in transit and in their workplaces and communities. The EEED research portfolio addresses environmental quality and regulation, energy resources and systems, water resources and systems, climate, natural hazards and disasters, and economic development—both domestically and internationally. EEED research is conducted for government, foundations, and the private sector.

Questions or comments about this monograph should be sent to the project leaders, Obaid Younossi (Obaid_Younossi@rand.org) and Nidhi Kalra (Nidhi_Kalra@rand.org). Information about EEED is available online (http://www.rand.org/ise/environ). Inquiries about EEED projects should be sent to the following address:

Keith Crane, Director
Environment, Energy, and Economic Development Program, ISE
RAND Corporation
1200 South Hayes Street
Arlington, VA 22202-5050
Tel.: 703-413-1100, x5520
Keith_Crane@rand.org

The RAND-Qatar Policy Institute

To study some of the most important issues facing the Middle East, RAND and the Qatar Foundation for Education, Science and Community Development formed a partnership that in 2003 established RQPI in Doha, Qatar. RQPI is an integral part of Education City, which is being developed by QF under the leadership of Her Highness Sheikha Mozah Bint Nasser Al Missned. Education City is a community of institutions—both K–12 and universities—contributing to education and research in Qatar and the Gulf region. RQPI is a regional office that facilitates delivery of the full range of RAND's capabilities to clients in North Africa, the Middle East, and South Asia—roughly, from Mauritania to Bangladesh.

Further Information
For further information about this report, other RQPI work on the Qatar National Research Fund, or RQPI, contact the director:

Obaid Younossi, Director
RAND-Qatar Policy Institute
P.O. Box 23644
Doha, Qatar
Tel.: 00974-4454-2500/02
Obaid_Younossi@rand.org

Contents

Figure and Tables

Figure

Tables

Summary

Purpose

Qatar is developing and growing rapidly. Its rich fossil fuel resources have enabled high economic growth in the past several decades. Today, it has one of the highest gross domestic products (GDPs) based on purchasing power parity per capita in the world.[1] Qatar also has a rapidly growing population.[2] Qatar has invested heavily in developing its national infrastructure, human capital, and social and cultural institutions.

Simultaneously, population growth and industrial development have increased Qatar's energy, water, and other resource needs substantially and threaten its environmental sustainability. In response, Qatar's leadership has created a vision of sustainable development for the country.[3] QF is moving this vision of sustainable development forward in part by establishing a national research institute that conducts and collaborates on applied research in energy, environment, and water issues.[4]

---

[1] International Monetary Fund, 2010.

[2] In 1960, Qatar's population was estimated to be only 45,000 people. By 2009, its population had grown more than 3,000 percent, to 1.4 million. This growth partly reflects a large influx of expatriates into the country (World Bank, undated).

[3] General Secretariat for Development Planning, undated (b).

[4] The Qatar Foundation is a private, chartered, nonprofit organization that addresses education, scientific research, and community development in the country (QF, undated).

To arrive at a better design for the Qatar Environment and Energy Institute, QF asked RQPI to recommend research priorities based on Qatar's needs and goals in these areas. These research priorities will help QF develop a research portfolio for the institute and frame an approach for shaping that portfolio over time.

These profiles will help QF identify areas in which there might be gaps or duplication in research areas, and the profiles would form the basis for exploring collaboration opportunities. This monograph presents the methodology and findings of our work.

Approach

This research was undertaken in two phases. In phase 1 of the project, RQPI surveyed the Gulf countries and existing literature to identify major research institutions in the GCC member states, areas of research on energy and the environment being conducted or planned by these institutions, and research programs funded by the GCC states undertaken outside of the GCC.

Phase 2 identified priority research areas for Qatar, focusing on the most important and promising in light of Qatar's unique challenges and resources. It drew on phase 1 work to identify opportunities for collaboration in the GCC and beyond in each of these areas.

Recognizing that there is a large and growing body of research in this area, as part of this endeavor, QF asked RQPI to survey the Gulf region to profile institutions engaged in energy and environment research. We undertook phase 2 in six steps and drew heavily on different expert panels, operating independently of each other, from RAND and Carnegie Mellon University (CMU). Steps 1–3 laid the groundwork for identifying priority research topics. In step 1, we generated a collection of research topics that could be considered for Qatar's institute. In step 2, we identified Qatar's energy, environment, and water situation, including goals and challenges. In step 3, we established the criteria by which topics should be assessed and selected. In steps 4 and 5, we brought these three elements together to select 14 priority research topics for the institute and synthesize our findings. In step 6, we reported on each priority research topic in a detailed white paper.

Findings on Environment and Energy Research in the Gulf Cooperation Council

We identified 15 major institutions in Qatar and the rest of the GCC member states directly engaged in research in energy and the environment. We determined the energy and environment research these institutions are undertaking or planning to undertake. Additionally, growing populations have sharply increased energy demands, making renewable energy sources and energy-efficiency key concerns. Where possible, we assessed missions, the extent to which they collaborated with each other, and other institutional features. We have used these observations to provide guidance for the institute.

Gulf Cooperation Council Countries Share Common Resources and Common Concerns

Countries in the GCC have common resources and concerns that are a result of shared geography, culture, and development. Common natural resources among GCC countries include sunlight, water and fish from the Gulf, crude oil, and natural gas. Sustainability is a shared concern among all GCC countries. The desert climate and regional geology limit the natural supply of freshwater and domestic capacity for food production, making the countries dependent on desalination. They also share a concern for protecting coastal resources from pollution to sustain fish stocks as a food supply. Additionally, growing populations have sharply increased energy demands, making renewable energy sources and efficient energy key concerns. Desert climates, climate change, and expanding populations also raise environmental concerns, such as desertification, which further reduces agricultural output. Environmental health, particularly air quality, is a concern due to rapid industrial expansion and high levels of particulate matter in the air.

Gulf Cooperation Council Institutions Are Engaged in a Range of Research Activities to Address These Concerns

The institutions we reviewed engage in a range of energy, water, and environment research to address these challenging issues. There are many common areas of research across countries and institutions. Common energy research includes developing solar energy, improving oil and natural gas recovery and reservoir management, developing carbon capture and storage (CCS), and using energy more efficiently with green buildings and smart grids. Water research largely focuses on desalination, increasing water efficiency, and wastewater management. Environmental research addresses air quality, marine and coastal environments, and conservation and rehabilitation. The research in the region addresses these issues through technology development, policy analysis, and resource management. Activities also range from basic science to applied research to commercialization of technologies. Also, many of the institutions we surveyed have a broader mandate than managing a research portfolio. Some are also involved in funding and policymaking.

Like Others, Qatar's Institute Should Take a Multidisciplinary Approach

Real-world energy and environment challenges are multidisciplinary. Accordingly, many of the institutions that we contacted take a multidisciplinary approach to research and study through multidisciplinary degree programs or integrated centers of excellence. We recommend that the institute similarly take an intentionally multidisciplinary approach to its research and focus its work around problem statements or topical centers of excellence. Additionally, there might be opportunities for the institute to collaborate with QF's two other research centers: the Qatar Biomedical Research Institute and the Qatar Computing Research Institute. In Chapter Seven, we suggest topics for which such cross-institutional collaboration might be appropriate.

Common Research Interests and a Shortage of Human Capital Suggest That the Institute Should Collaborate

The common research interests among GCC institutions leads to some duplication of research. Duplication in critical areas of research could be desirable and lead to faster, better, or more-diverse solutions to energy and environment problems. Although there is a desire for collaboration in the region, actual collaborative efforts between institutions in the GCC and even within countries have been limited. It could also help to build local human capital, which many regard as the biggest hurdle to achieving regional research objectives.

Common research interests and a shortage of human capital suggest a great need for collaboration and coordination among GCC institutions, which could also increase the capacity for quality research. Although there is a desire for collaboration in the region, actual collaborative efforts between institutions within the GCC and even within the same country have been limited.

We recommend that the Qatar Environment and Energy Institute develop relationships with institutions and other scientists in the region. We suggest that QF draw from the GCC Energy and Environment Research Database we have developed to identify potential institutional partners. We also urge QF to develop the institute's research portfolio and programs in close coordination with other institutions in the country and collaborate where possible. This will help avoid duplicate research where it is not needed. Sharing human capital can help avoid competition between institutions, which could be counterproductive to addressing research problems. Along with shared facilities, this can also help the institute undertake research quickly.

The Institute Should Seek Input from and Involve the Qatari Government

Government funding and policymaking are needed to promote research that can have long-term benefits and promote the public good. Additionally, government is a key stakeholder and a potential collaborator for research, particularly in public policy research. Thus, it is important that research institutions and governments align their research priorities and collaborate. In our survey, we found that, although GCC governments are involved in funding research on energy and the environment, their funding priorities are not always clearly identified or integrated with existing research capabilities.

We recommend that QF seek input from and involve the Qatari government. The Qatari government recently articulated some of its energy and environment priorities in the launch of its *National Development Strategy* for 2011–2016.[5] The institute's overall research portfolio should be consistent with these and other priorities. Additionally, the institute should engage with government agencies, such as the Ministry of Environment and the Ministry of Energy and Industry, and others to ensure that

- research priorities are aligned in specific areas
- resources allocated to energy and environment research reflect the institute's capabilities and needs
- research findings reach their intended audience
- the institute and government agencies collaborate.

Recommendations for Priority Research Topics

We identified 14 priority research topics that QF should consider for the Environment and Energy Institute (see Table S.1). These topics address Qatar's most-pressing issues in energy production and use, water production and use, and environmental preservation.

Recommended Topics for Energy Research

Oil and natural gas are Qatar's critical energy resources, and they are the backbone of Qatar's economy and its critical industries. Today, Qatar faces important opportunities and challenges in oil and natural gas. Qatar's oil and gas are limited resources, and Qatar must find ways of using these resources efficiently and to the greatest benefit for its economy and population. Fossil fuels such as these are responsible for approximately 75 percent of global anthropogenic (human-

---

5   General Secretariat for Development Planning, undated (a).

Table S.1
Priority Research Topics for Qatar's Environment and Energy Institute

Topic Area	Topic
Energy	1. Natural gas production and processing
	2. Petroleum production and processing
	3. CCS
	4. Solar energy development
	5. Fuel cells
	6. Green buildings
	7. Smart grids
	8. Strategic energy planning
Water	9. Desalination
	10. Groundwater sustainability
	11. Water demand management
	12. IWRM
Environment	13. Environmental characterization
	14. Crosscutting environmental research

NOTE: Numbers correspond to topic numbers within this book and are not intended to convey any sense of priority among the topics. IWRM = integrated water resource management .

caused) greenhouse gas (GHG) emissions from the past 20 years. Thus, although fossil fuels remain the primary source of energy globally, demands for clean energy—including clean fossil fuels—are growing. Qatar is seeking ways of responding to these demands. Additionally, as Qatar's fossil fuel reserves are depleted, Qatar will need alternative, renewable forms of energy.

We recommend that the institute help Qatar address these challenges and capitalize on opportunities by conducting research in several key areas. The institute should improve the production and processing of natural gas and petroleum and help Qatar use CCS to respond to the demand for clean energy. It should develop solar energy to conserve these resources and provide an alternative, renewable source of energy, and develop fuel cells to improve electricity generation in the long term. The institute should also conduct research that increases domestic energy efficiency through the use of green buildings and smart grids. Finally, the institute should help Qatar develop a long-term strategic energy plan to help prioritize competing energy issues.

We recommend that the institute conduct research to improve Qatar's *natural gas production and processing* (topic 1). Natural gas has several advantages over other fossil fuels: It burns more cleanly and emits fewer GHGs than other fossil fuels, and natural gas–fired power plants are less expensive to construct and operate than coal-fired power. Because of this, natural gas is already a preferred fuel for electricity generation in the United States and elsewhere. Yet, there are resource, logistical, economic, and environmental challenges to the increased development of natural gas in Qatar. Thus, research should focus on two strategic issues: (1) minimizing emissions of GHGs from natural gas and its derivative products and (2) meeting the growing demand for specialized refined and synthetic products and on improving liquefied natural gas operations.

We also recommend that the institute conduct research to improve *petroleum production and processing* (topic 2). Qatar Petroleum spearheads petroleum research in Qatar. Its research needs and agenda are not public, but it is known that Qatar Petroleum is undertaking research in a wide range of areas, including improving oil and gas operations, environment and climate change, and reservoir engineering and management. We believe that reservoir engineering and management, which includes estimating reserve sizes and developing extraction strategies for high recovery rates, are particularly important for Qatar. The institute should directly and closely collaborate with Qatar Petroleum to identify a complementary research agenda and to execute its research. The institute can play an important role in contributing to long-term, environmentally conscious development and use of Qatar's petroleum resources.

The institute should also research *CCS* (topic 3), the process of capturing carbon dioxide (CO_2) and depositing it in a reservoir for permanent storage. This prevents CO_2 from contributing to the GHG effect and makes fossil fuel use more environmentally sustainable. The institute's research should focus on two aspects of CCS most relevant to Qatar. First, research should address the technical and logistical challenges of storing CO_2 in the region, including identifying storage sites and assessing both their long- and short-term stability for containing CO_2. Second, the institute should initiate research on developing natural gas combustion–specific CCS technologies.

Solar energy (topic 4) is a largely untapped resource for Qatar and is its only major source of renewable energy. We recommend that the institute conduct research in four areas: designing hybrid power plants with solar energy and natural gas, advancing solar energy–driven space and water heating and cooling, advancing solar photovoltaics (PV), and improving solar desalination.

Fuel cells (topic 5) are electrochemical devices that combine a fuel (such as natural gas) and oxygen to produce electricity. Fuel cells are more efficient than combustion-based methods, using less fuel and emitting fewer GHGs per kilowatt-hour. They use much less water, and, if used to generate electricity at or near the use site, they can reduce or eliminate transmission losses. We recommend that the institute conduct research to reduce the cost of fuel cells and increase their power output without sacrificing their reliability or longevity.

Improving energy sources is an important component of the institute's research agenda, but greater efficiency is also an important research area. We recommend that the institute undertake green building and smart grid research to improve energy efficiency.

Green buildings (topic 6)—buildings that use energy, water, and natural resources efficiently—can help Qatar's economic development become more sustainable. We recommend that the institute conduct research on improving green building designs and technologies with particular emphasis on water and energy conservation. This research should also consider effectiveness, feasibility, and appropriateness of different policy options to encourage the use of green building designs and technologies.

Smart grids (topic 7) consist of flow-control and monitoring technologies that can improve efficiency, reliability, security, and flexibility of the electrical grid. They offer several potential advantages for Qatar:

more-efficient use of the gas supply, integration of distributed renewable energy sources, and pricing mechanisms that can reduce demand. The institute should pursue smart grid research if such benefits become key energy goals for Qatar.

Many activities in the energy sector are interlinked. For example, an emphasis on solar PV makes smart grids an important area to pursue. *Strategic energy planning* (topic 8) is needed to ensure that all elements operate in concert and not at cross-purposes. This strategy should establish national priorities and take into account the linkages between the energy sector and other parts of the economy.

Recommended Topics for Water Research

Water is a critical resource, and water security is a critical issue in Qatar. Water consumption is high, for both municipal and industrial use. We recommend three areas of water research: improving water supplies through desalination and groundwater sustainability, managing water demand, and developing a holistic process for achieving water security through IWRM.

Qatar uses *desalination* (topic 9) to produce about 180 billion liters of water annually. Its desalination plants are energy-intensive and burn fossil fuels. The country should explore technologies that improve the efficiency of the process and drive down the cost. We recommend that the institute conduct research on technologies to increase the energy efficiency of thermal desalination, including hybridizing the thermal desalination process with reverse osmosis desalination, a process that requires much less energy than other methods do. We also recommend that the institute conduct research on alternative methods of desalination, improved materials and processes for desalination, and colocation of desalination and power plants.

Qatar's *groundwater* (topic 10) is overexploited. One potential result of this exploitation is that groundwater could be depleted in the coming years. Qatar needs a comprehensive groundwater sustainability research program to help manage this critical resource. A first step would be a comprehensive monitoring program. Complementary research could focus on reducing demand for groundwater, reducing demand for water services, developing alternative options for agriculture water services, and developing ways to increase the amount of groundwater. The institute should also research how climate change could affect Qatar's groundwater.

Key to the issue of water supply is *managing demand* (topic 11). The institute should undertake a comprehensive research agenda to reduce the consumption of water. Research on water demand-side management should include pricing and nonprice policies for residential water and agricultural demand for water, reducing transmission and distribution losses, and developing education campaigns to encourage conservation.

Because the approaches to deal with water management are interrelated, Qatar should *integrate water resource management* (topic 12). IWRM is a holistic approach to water management that takes into account links between different types of water sources, as well as the links between water management and other aspects of the economy. We recommend that the institute help Qatar assess and potentially adopt an IWRM process that is attuned to Qatar's needs. If the institute determines that IWRM would be valuable and the Qatari government agrees to embark on an IWRM process, then the institute should help identify an appropriate framework for implementing IWRM in Qatar.

Recommended Topics for Research on the Environment

The combined forces of population growth, industrialization, and coastal development have adversely affected Qatar's environment. Sustainable development indicators from Qatar's Statistics Authority suggest that air pollution levels in Doha are rising, that many land areas are undergoing desertification, and that Qatar's marine environment is damaged. Qatar has taken several steps to address its environmental challenges. However, a great deal of uncertainty remains about the current health of Qatar's environment.

We recommend that the institute first *characterize the current state of Qatar's environment* (topic 13). This includes coastal and desert ecosystem and biodiversity characterization. It also includes assessing human-made pollution and waste in the air, water, and land. The study of ecosystems and pollutants are linked: For example, coastal water pollution will affect the health of fisheries and coastal ecosystems.

We also recommend that the institute undertake *crosscutting environmental research* (topic 14) that will help Qatar address these environmental challenges and that will inform water and energy issues. This includes evaluating risks from climate change; using environmental economics to identify solutions to energy, water, and environment challenges; and undertaking a mass–energy balance.

Next Steps

These priorities are only the beginning of a journey to realize such a bold vision. We recommend that QF develop an initial research portfolio for the institute from the list of priority topics identified in this research.

Select Initial Research Topics

The institute's portfolio should consist of only two or three topic areas in the initial two to three years. This controlled growth allows some diversity in research capabilities from the beginning, while also enabling the institute to mature and develop a governance approach, funding mechanisms, an institutional culture, partnerships, and other facets. The two to three initial research topics will serve as a test bed for this institutional framework.

Initial topics should be of utmost importance to Qatar's energy and environment and ones in which research results can be applied in the near term. The choice of topic will also depend on the ability to recruit research staff and develop facilities, whether the work is complementary to other work in Qatar or the region, and whether Qatar has a strategic advantage in undertaking it.

This monograph principally offers guidance on the importance of a topic to Qatar's energy and environment goals, and it offers insights about human capital and about research being conducted in the region. Based on this, several topics stand out as strong candidates for these initial topics. These are natural gas (topic 1), desalination (topic 9), groundwater (topic 10), water demand–management research (topic 11), and environmental characterization (topic 13).

Create Conditions for Continued Growth

QF should also create conditions for this research portfolio to grow and address many of the 14 priority research areas. This can be done in the near term by developing a five- to ten-year roadmap for expanding the institute's research capabilities into other areas. This top-down mapping can be complemented by a bottom-up approach to growth, in which investigators suggest research topics, submit proposals, and compete for supplementary funding from the institute. Such an approach draws on the talents and knowledge of existing researchers, creates opportunities for more junior staff to lead research, can serve as a form of peer review and quality assurance, and grows the program organically. Because it requires a modest base of researchers, this mechanism might be feasible only after the first few years.

Track Progress

This endeavor is ambitious. It will require a long-term commitment by QF. As we recommend for the research it will be sponsoring with the institute, we recommend that QF approach this overall endeavor as an experiment in itself, in which decisions to invest in a research area are treated as hypotheses that can be tested and altered as new evidence is available. The hypothesis is that the recommended research topic will yield benefits to Qatar at an investment cost that is reasonable. This hypothesis can be testable by estimating potential outcomes and comparing them with evidence that emerges as research is funded and undertaken. We recommend that the institute focus on at least two kinds of outcomes that reflect the institute's mission: (1) the extent to which the institute is furthering Qatar's energy and environment goals and (2) the extent to which it helps realize Qatar's goals of becoming an international leader in research. Initial topics, metrics, and thresholds should be developed as part of the mapping and strategic-planning process, but the slate of research topics presented here provides seeds for this endeavor to grow.

Acknowledgments

We would like to express our deep appreciation to our sponsor, the QF research division. In particular, we are grateful to Abdelali Haoudi, vice president of the QF research division, for sponsoring the project. We thank Dirar Khoury, director of institutional research for the research division, and Chee Wen Chong for their constant support and guidance throughout this effort. We thank Rabi Mohtar, director of the Qatar Environment and Energy Institute, for his review of this work.

We would further like to acknowledge the valuable insights and observations contributed by our colleagues James T. Bartis, Henry H. Willis, Scott Hassell, David G. Groves, Jordan R. Fischbach, Susan Bohandy, Jerry M. Sollinger, and Keith Crane. We greatly appreciate Nicholas Burger's thoughtful review of and recommendations for the monograph. We are also grateful for the hard work of Sarah Hauer, Jennifer Miller, and Lisa Bernard for compiling, formatting, and producing the manuscript.

We thank our colleagues at Carnegie Mellon University for their generous time serving on expert panels and providing key input into this research: M. Granger Morgan, Inês Azevedo, David Dzombak, Kelvin Gregory, Gabriela Hug, Paulina Jaramillo, Krishnapuram Karthikeyan, Sean T. McCoy, and Francis C. McMichael. We also thank colleagues at the National Renewable Energy Laboratory (NREL), the Joint Institute for Strategic Energy Analysis (JISEA), and the University of Colorado Boulder for their insights on developing research institutions in energy and environment: Douglas Arent (NREL/JISEA), who also provided an insightful review of the monograph; Robert Baldwin (NREL); Scott Huffman (NREL); David S. Ginley (NREL); and Carl A. Koval (University of Colorado Boulder).

The GCC survey research in particular could not have been accomplished without the assistance of many individuals across the Gulf region who supplied us with data, information, and expertise. We are grateful to Abdul Sattar Al-Taie at Qatar National Research Fund for supplying us with grant information for funded projects within Qatar. We would also like to recognize Yousif Al Hamar and Salman Salman from the Ministry of Environment, Omran Al-Kuwari from GreenGulf, Carl Atallah of Chevron Qatar Energy Technology, Daniel Ramadan from Qatar Science and Technology Park, Ahmed Abdel-Wahab and Patrick Linke at Texas A&M University at Qatar, Majeda Khraisheh and Hassan Rashid Al Derham at Qatar University, and Samer Adham from ConocoPhillips Global Water Sustainability Center. At United Arab Emirates University in Al Ain, we would like to thank Wyatt R. (Rory) Hume, Donald E. Bowen, Reyadh Almehaideb, Eyad H. Abed, and M. Naim Anwar for sharing project information and their personal expertise in environment and energy research. We are grateful to Wajih N. Sawaya and Naji Al-Mutairi from the Kuwait Institute for Scientific Research for sharing their time and expertise. From Masdar Institute of Science and Technology in Abu Dhabi, we would like to thank John Perkins and Marwan Khraisheh for sharing their time and insights. In Oman, we benefited from the assistance of Shannon McCarthy at the Middle East Desalination Research Center and Mushtaque Ahmed, Rashid S. Al-Maamari, and Bassam Soussi at Sultan Qaboos University. Finally, we would like to thank Mohamed Noman Galal and Abdulla M. Al-Sadiq at the Bahrain Center for Studies and Research for their assistance and hospitality.

Abbreviations

AUST	Ajman University of Science and Technology
BCSIES	Bahrain Centre for Strategic, International and Energy Studies
BCSR	Bahrain Center for Studies and Research
BIPV	building-integrated photovoltaics
BQDRI	Barwa and Qatari Diar Research Institute
CCS	carbon capture and storage
CCUS	carbon capture, use, and storage
CEMB	Center of Excellence in Marine Biotechnology
CEO	chief executive officer
CERT	Center of Excellence for Applied Research and Training
CESAR	Center for Environmental Studies and Research
CHP	combined heat and power
CMU	Carnegie Mellon University
CO_2	carbon dioxide
CPV	concentrating photovoltaics
CSP	concentrating solar power

DAWR	Department of Agriculture and Water Research
DSM	demand-side management
DWR	Department of Water Resources
EEED	Environment, Energy, and Economic Development Program
EMEC	European Marine Energy Centre
EOR	enhanced oil recovery
EPA	U.S. Environmental Protection Agency
ERC	energy resource characterization
GCC	Gulf Cooperation Council (formally, the Cooperation Council for the Arab States of the Gulf)
GDP	gross domestic product
GESR	Gulf Electronic Scientific Research
GHG	greenhouse gas
GIS	geographic information system
GTL	gas to liquid
GWP	Global Water Partnership
GWSC	Global Water Sustainability Center
IGCC	integrated gasification combined cycle
IP	intellectual property
ISCCS	integrated solar combined-cycle system
ISE	RAND Infrastructure, Safety, and Environment
ISO	International Organization for Standardization
IT	information technology

IWRM	integrated water resource management
JCCP	Japan Cooperation Center, Petroleum
JISEA	Joint Institute for Strategic Energy Analysis
KACST	King Abdulaziz City for Science and Technology
KAUST	King Abdullah University of Science and Technology
KFUPM	King Fahd University of Petroleum and Minerals
KISR	Kuwait Institute for Scientific Research
kj	kilojoule
KSA	Kingdom of Saudi Arabia
LNG	liquefied natural gas
LRMC	long-run marginal cost
LTTD	low-temperature thermal desalination
m^3	cubic meter
M&V	monitoring and verifying
MEA	Millennium Ecosystem Assessment
MED	multieffect distillation
MEDRC	Middle East Desalination Research Center
MENA	Middle East and North Africa
MIT	Massachusetts Institute of Technology
mmBtu	million British thermal units
MOU	memorandum of understanding
MSF	multistage flash

MW	megawatt
NGCC	natural gas combined cycle
NOx	nitrogen oxide
NPRP	National Priorities Research Program
NREL	National Renewable Energy Laboratory
O&M	operation and maintenance
OAR	Office of Academic Research
OECD	Organisation for Economic Co-Operation and Development
OGRC	Oil and Gas Research Center
PEM	polymer electrolyte membrane
PI	principal investigator
PP	polypropylene
PV	photovoltaics
QF	Qatar Foundation for Education, Science, and Community Development
QNFSP	Qatar National Food Security Programme
QNRF	Qatar National Research Fund
QSAS	Qatar Sustainability Assessment System
QSTP	Qatar Science and Technology Park
QU	Qatar University
QWE	Qatar Sustainable Water and Energy Utilization Initiative
REC	renewable energy credit
RO	reverse osmosis

RQPI	RAND-Qatar Policy Institute
SABIC	Saudi Basic Industries Corporation
Saudi Aramco	Saudi Arabian Oil Company
SCENR	Supreme Council for the Environment and Natural Reserves
SEGS	solar-electric generating station
SOFC	solid-oxide fuel cells
SQU	Sultan Qaboos University
SRES	Special Report on Emissions Scenarios
TcF	trillion cubic feet
TERI	Energy and Resources Institute
UAE	United Arab Emirates
UAEU	United Arab Emirates University
UCS	Union of Concerned Scientists
UNCCD	United Nations Convention to Combat Desertification
UNDP	United Nations Development Programme
UREP	Undergraduate Research Experience Program
VCUQatar	Virginia Commonwealth University in Qatar
WCRP	World Climate Research Programme
WRC	Water Research Center
YSZ	yttria-stabilized zirconia

CHAPTER ONE

Introduction

Background

Qatar is developing and growing rapidly. Its rich fossil fuel resources have enabled high economic growth during the past several decades. Today, it has one of the highest gross domestic products (GDPs) based on purchasing-power parity per capita in the world.[1] Qatar also has a rapidly growing population.[2] Qatar has invested heavily in developing its national infrastructure, human capital, and social and cultural institutions.

But there have been drawbacks to Qatar's rapid growth as well. Population growth and industrial development have increased Qatar's energy, water, and other resource needs substantially and threaten its environmental sustainability. Its economy is also largely based on oil and natural gas, which are limited resources. The leaders of Qatar have long recognized these problems and have articulated a vision of Qatar developing sustainably and evolving into a knowledge-based economy. *Qatar National Vision 2030* is a specific statement of several long-term goals for the country as a whole.[3] Some of the goals described in this vision statement address sustainable development and economic diversity issues, such as

---

[1] International Monetary Fund, 2010.

[2] In 1960, Qatar's population was estimated to be only 45,000 people. By 2009, its population had grown more than 3,000 percent, to 1.4 million. This growth partly reflects a large influx of expatriates into the country (World Bank, undated).

[3] General Secretariat for Development Planning, 2009.

- balancing development with environmental protection
- responsibly exploiting oil and gas
- realizing competitive advantages in industries derived from oil and natural gas
- evolving into a diversified, knowledge-based economy characterized, in part, by innovation and entrepreneurship.[4]

It also calls for "[e]ffective and sophisticated environmental institutions that build and strengthen public awareness about environmental protection, and encourage the use of environmentally sound technologies."[5]

The Qatar Foundation for Education, Science, and Community Development (QF) is moving this vision forward in part by establishing a national research institute that conducts applied research on the country's most-pressing energy, environment, and water issues.[6]

Purpose

To arrive at a better design for the Qatar Environment and Energy Institute at its inception, QF asked the RAND-Qatar Policy Institute (RQPI) to recommend research priorities based on Qatar's needs and goals in these areas. These research priorities will help QF develop a research portfolio for the institute and frame an approach for shaping that portfolio over time.

Recognizing that there is a large and growing body of research in this area, as part of this endeavor, QF asked RQPI to survey the Gulf region to profile institutions engaged in energy and environment research. These profiles will help QF identify areas in which there might be gaps or duplication in research areas, and they would form

---

[4] These goals are closely paraphrased from the *Qatar National Vision 2030* statement.

[5] General Secretariat for Development Planning, 2009, p. 8.

[6] QF is a private, chartered, nonprofit organization that addresses education, scientific research, and community development in the country ([QF], undated).

the basis for exploring collaboration opportunities. This monograph presents the methodology and findings of our work.

How the Monograph Is Organized

The monograph has eight chapters. Chapter Two describes the methodology used to scan research in the Cooperation Council for the Arab States of the Gulf (also known as the Gulf Cooperation Council, or GCC) and to identify the key research topics. Chapter Three presents findings about energy and environment research in the GCC. Chapters Four through Six present the 14 priority topics identified in the three research categories: energy (Chapter Four), water (Chapter Five), and environment (Chapter Six). Chapter Seven discusses research topics identified during the topic-development process that might be important and promising but not critical. Chapter Eight outlines the next steps QF should take. The monograph also has five appendixes. Appendix A lists potential research topics and subtopics, and Appendix B presents interim recommendations from expert panels. Appendix C provides profiles of each institution reviewed in the survey of GCC research. Appendix D provides the interview protocol for the GCC survey, and Appendix E gives the search terms we used for our literature review.

CHAPTER TWO

Project Methodology

Overview

This project was aimed at identifying priority research topics that QF should consider for the Environment and Energy Institute's research program. Our recommendations are based on an analysis of how well different research topics serve the mission of helping Qatar develop sustainably through applied research in technology, policy, and management. This monograph also provides a survey of other research being undertaken in the region.

This research was undertaken in two phases. In phase 1 of the project, RQPI surveyed the Gulf countries and existing literature to identify major research institutions in the GCC member states, areas of research on energy and the environment being conducted or planned by these institutions, and research programs funded by the GCC states undertaken outside of the GCC.

Phase 2 identified priority research areas for Qatar, focusing on the most important and promising in light of Qatar's unique challenges and resources. It drew on phase 1 work to identify opportunities for collaboration in the GCC and beyond in each of these areas.

Methodology for Assessing Gulf Cooperation Council Environment and Energy Research (Phase 1)

To assist QF in better targeting its resources, we surveyed institutions in the GCC by conducting semistructured interviews over the telephone and during site visits. These institutions included universities, laboratories, and industry. To capture important regional themes and priorities, we also reviewed publications discussing research in the GCC and the broader Middle East and North Africa (MENA). We assembled this information into the GCC Energy and Environment Research Database, which we have provided to QF.

Survey of Institutions

In the first phase of the project, we gathered information on the institutions that have been conducting research on energy and the environment and their research profiles. We identified 28 major institutions within Qatar and the GCC engaged in research in energy and the environment.[1] These institutions were identified based on recommendations from subject-matter experts and literature reviews. During this process, researchers also identified six additional institutions within the MENA region that are also conducting research in these areas.

We developed a written interview protocol and survey instrument to obtain information on the most-prominent past and ongoing research conducted by these institutions in the areas of energy and environment, as well as their future plans. This instrument can be found in Appendix D. The information collected from the survey included institutional information regarding organizational structure, mission, and partner institutions. Additional information about the projects and programs included sources of funding, cost of projects, and methodologies employed. The survey instrument was first tested in Qatar and the United Arab Emirates (UAE) through semistructured interviews and then used in interviews or emailed to other GCC institutions identified from the list of major institutions. In addition to enabling us to gather information about the institutions, the interviews enabled us to draw on the expertise of energy and environment researchers to identify promising research areas for Qatar's new institute.

---

[1]   We would like to note that we sought to identify the most-prominent and most-active institutions in the region based on publicly available information. There might be other institutes engaged in this type of research that we were unable to identify, particularly within the private sector and the oil and gas industry, as their research and activities are not always public.

Interviews and Site Visits

We conducted 14 site visits across Qatar and the GCC and held interviews and discussions with more than 30 participants, including university and institution leadership, department heads and division leads, government officials, industry representatives, and key researchers in energy and the environment. We initially identified 12 institutions outside Qatar for these site visits, based on their research focus and relevance to the institute. We received invitations for visits from six of the 12 institutions. We used the same interview protocol for each semistructured interview. This protocol is provided in Appendix D. Most of the institutions we visited provided us with materials, such as annual research reports, which included details on the types of ongoing projects at the institution. Some institutions prepared presentations or tailored materials about their environment and energy research. To ensure that we had accurately captured all relevant data on each institution and its research portfolio, we also cross-referenced the information provided to us at meetings with open-source information on individual institutions.

In Qatar, we conducted semistructured interviews with representatives from industry, government, and academia, including the following:

- Qatar National Research Fund (QNRF)
- Qatar Science and Technology Park (QSTP)
- Ministry of Environment
- ConocoPhillips
- Chevron
- GreenGulf
- Texas A&M University
- Qatar University.

Outside Qatar, we conducted site visits and interviews with senior managers and key researchers from various institutions, including the following:

- United Arab Emirates University (UAEU)

- Masdar Institute of Science and Technology, UAE
- Bahrain Center for Studies and Research (BCSR)
- Kuwait Institute for Scientific Research (KISR)
- Middle East Desalination Research Center (MEDRC), Oman
- Sultan Qaboos University (SQU), Oman.

Literature Review

In parallel with the survey and site visits, we reviewed published literature to determine the dominant themes of research related to energy and the environment, delineate these themes by country within our region of interest, and assess the prospects and maturity of projects addressed in the literature that relate to these themes. We developed a map of research efforts and projects that offered potential solutions to leading issues in energy and the environment, identified the leading figures and institutions behind these projects, and drew the inter- and intranational links developed for these projects.

To do this, we first laid out major themes and fields of research related to energy and the environment in the GCC and MENA based on initial discussions with the client and basic information gathered on institutions in the region.[2] We divided research into three main themes: water, energy, and the environment. Although these are clearly interrelated topics, these are the three main focus areas that emerged in our initial review of institutions. For instance, SQU in Oman has three core research centers of excellence for energy and environment: the Water Research Center (WRC), the Center for Environmental Studies and Research (CESAR), and the Oil and Gas Research Center (OGRC). We identified 16 topics within these three areas that were relevant to the GCC and MENA.[3] A list of these topics is provided in Table 2.1.

---

[2]   Although the survey, interview, and site-visit portion of the study focused on the GCC, we expanded the literature review to include the entire MENA region in order to identify research priorities in the broader region that might influence GCC research activities.

[3]   This list was expanded and refined in task 2 of the project based on findings from the survey and inputs from experts.

Table 2.1
Initial Topic List

Topic Area	Topic
Energy	1. Solar
	2. Wind
	3. Fuel cells
	4. Smart grids
	5. Green building
	6. Efficient oil and gas production
	7. Energy impact on water
	8. GTL
Environment	9. GHG emissions and carbon sequestration
	10. Air pollution
	11. Environmental effects of oil and gas production
Water	12. Water use and wastewater
	13. Water management
	14. Novel methods of producing water
	15. Water recycling and reuse
	16. Impact on energy efficiency

NOTE: Numbers correspond to topic numbers in this book and are not intended to convey any sense of priority among the topics. GHG = greenhouse gas. GTL = gas to liquid.

Next, we used the list of institutions and countries and the 16 topic areas to generate more-specific search terms for finding and reviewing published material related to energy and environment in the GCC and MENA. We reviewed reports, books, and journals from institutions in more than 15 countries in MENA, Central Asia, North America, and Europe. We also drew on articles discussing research from Qatari public media, such as *The Peninsula* and *Gulf Times*, as well as global media sources, such as *The New York Times*. Last, we examined RAND publications related to energy or the environment.

Appendix E lists the databases searched, the search terms used, and the sources of publications.

By drawing research from several databases, we made certain that we were collecting information from a range of government and non-government institutions, academic researchers, and the private sector. A complete list of the relevant literature with citations can be found in the GCC Energy and Environment Research Database that was provided to QF.

We also identified examples of regional and international partnerships that have been developed between organizations and institutions to conduct research on energy and the environment. These partnerships can be classified into three sets. The first set of partnerships consists of governmental and institutional public-sector relationships. The second set of partnerships is based on academia and a more–academically focused exchange between countries and institutions. The third set of partnerships is based on commerce and investment prospects.

Gulf Cooperation Council Environment and Energy Research Database Development

We incorporated this information into an updatable database on research on energy and environment in the GCC. The GCC Energy and Environment Research Database draws from several sources, including the survey, tailored briefings to the RQPI team, QNRF's database, published journal articles, research reports, annual reports, institutional websites, and directories of researchers. The database includes a list of 28 institutions in the GCC and five additional institutions in the greater MENA region, 362 recent and ongoing projects within these institutions that are related to energy and environment topics, and a list of 59 recent publications with citations.[4] Table 2.2 shows a breakdown of the data collected by theme and type.

The research project information in the GCC Energy and Environment Research Database is limited by the amount and quality of information that we were able to collect from various sources because

---

[4] In some cases, complete research portfolios and project information were not available to us.

Table 2.2
Research Projects and Publications in the Gulf
Cooperation Council Energy and Environment
Database, by Theme

Key Thematic Areas	Research Projects	Publications
Water	140	12
Energy	121	37
Environment	101	10

some information is not publicly provided. Thus, the 362 projects listed in this database are not comprehensive but can be viewed as a representative sample of the types of research being conducted on energy and the environment in the region. The GCC Energy and Environment Research Database can be used to inform future collaboration decisions.

Methodology for Identifying Priority Research Topics (Phase 2)

We undertook phase 2 in six steps, as illustrated in Figure 2.1. Steps 1–3 laid the groundwork for identifying priority research topics. In step 1, we generated a collection of research topics that could be considered for Qatar's institute. In step 2, we identified Qatar's energy, environment, and water situation, including goals and challenges. In step 3, we established the criteria by which topics should be assessed and selected. In steps 4 and 5, we brought these three elements together to select priority research topics for the institute and synthesize our findings. In step 6, we reported on each priority research topic in a detailed white paper. These white papers are presented as priority topic subsections in Chapters Four through Six.

Use of Experts
Qatar faces many and complex sustainable development challenges. The research topics in energy, water, and the environment that could be undertaken to address those challenges are equally complex and

Figure 2.1
Priority Topics Were Selected Using a Six-Step Process

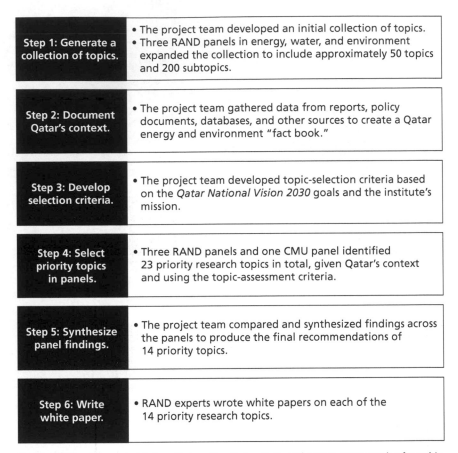

Step 1: Generate a collection of topics.	• The project team developed an initial collection of topics. • Three RAND panels in energy, water, and environment expanded the collection to include approximately 50 topics and 200 subtopics.
Step 2: Document Qatar's context.	• The project team gathered data from reports, policy documents, databases, and other sources to create a Qatar energy and environment "fact book."
Step 3: Develop selection criteria.	• The project team developed topic-selection criteria based on the *Qatar National Vision 2030* goals and the institute's mission.
Step 4: Select priority topics in panels.	• Three RAND panels and one CMU panel identified 23 priority research topics in total, given Qatar's context and using the topic-assessment criteria.
Step 5: Synthesize panel findings.	• The project team compared and synthesized findings across the panels to produce the final recommendations of 14 priority topics.
Step 6: Write white paper.	• RAND experts wrote white papers on each of the 14 priority research topics.

NOTE: CMU = Carnegie Mellon University. *Qatar National Vision 2030* can be found in General Secretariat for Development Planning, 2009.
RAND *MG1106-2.1*

numerous. Determining which topics should receive research attention requires a wide range of expert knowledge and judgment. Throughout this research project, we consulted 20 research experts: 11 from RAND and nine from CMU.

Both RAND and CMU have broad experience in science, engineering, and public policy, and both have experience and resident offices in Qatar. Many of the researchers we consulted had expertise in more than one of the three areas (energy, water, and the environment), and all were knowledgeable of crosscutting issues, such as climate change and the relationship between water consumption and energy production. Some also have prior experience in developing science and technology research priorities for research institutions.

We engaged experts in open-ended discussion panels rather than by means of rank ordering, Delphi, consensus scoring, or other highly structured methods. Discussions allowed open-ended debate on these complex issues and facilitated the creative thinking that was necessary for this project. This approach was also necessary in order for experts to work with a diverse body of unstructured background information. Importantly, it enabled experts to consider the interrelationships between research topics and to reorganize and refine topics at each step of the process, which could not be easily done by scoring or voting. This reorganization and restructuring of topics was particularly important in shaping a research agenda that was uniquely tailored to Qatar's needs.

Step 1: Develop a Collection of Potential Topics

We assembled a collection of potential research topics in energy, water, and the environment. These would seed the expert panels' subsequent assessments of topics (step 4), given Qatar's goals and challenges in these areas, and given the institute's mission.

We sought to be as inclusive as possible to ensure that the later assessment would not overlook key topics. Therefore, the project team first developed an initial list of potential research topics in each of the three areas, independent of Qatar's specific energy and environment challenges. We based this list on a broad literature review, the content of research programs around the world, and our team's knowledge of recent and current research. We focused on applied research topics and organized the topics into broad themes (e.g., energy sources and energy use). We did not limit our list to a particular number of topics.

We then consulted with RAND experts in three separate water, energy, and environment discussion panels. The experts reviewed and expanded the initial list of potential topics based on their own knowledge, again with the objective of being broadly inclusive. Experts also revised the organization of topics. In total, the expert review identified approximately 50 potential research topics.

Experts further divided the topics into specific subtopics that spanned three categories: technology development, policy analysis, and resource management. For example, the topic of groundwater sustainability included such research subtopics as developing nontraditional methods of groundwater recharge (technology development), researching pricing policies to manage groundwater demand (policy analysis), and monitoring and characterizing groundwater (resource management).[5] This collection of potential topics and subtopics is provided in Appendix A.

Step 2: Understand Qatar's Energy, Environment, and Water Situation

We next gathered extensive information about Qatar's energy, water, and environment sustainability goals and challenges. This included statistics about Qatar's current energy and water supply and demand, projections for the future, and key environmental challenges. Our sources included academic literature, reports, and data from Qatar's Statistics Authority; official government reports; reports from local industries and institutions (such as Qatar Petroleum) and international institutions (such as the World Bank). We also determined Qatar's policies and goals in these areas from such documents as *Qatar National Vision 2030*.[6] We supplemented published information with the knowledge of staff at RQPI. We assembled this information into an informal fact book, which experts used to inform their selection of research topics in step 4.

Step 3: Develop Criteria for Selecting Research Topics

We next established the criteria by which a research topic might be considered promising for the institute. A research program should

---

[5] Environment experts did not differentiate individual environment topics into subtopics. They instead recommended that every environmental research area could include research on sources, fate and transport, exposure, and treatment.

[6] General Secretariat for Development Planning, 2009.

be designed to further strategic objectives.[7] For private firms, for example, strategic objectives could include increasing the profitability of a particular product or projecting the firm's existing products into new markets.

Qatar's objective is to develop sustainably, balancing economic development with environmental protection, and to evolve into a diverse, knowledge-based economy. *Development* implies growth of industries, such as Qatar's oil and natural gas, growth of new industries, and an improvement in people's quality of life. *Sustainability* implies a conservation of resources (e.g., oil and natural gas reserves) and protection of environments and ecosystems (e.g., Qatar's fisheries, deserts, air, land, and the global climate). The institute's primary objective is to help Qatar develop sustainably by conducting applied research.

We derived four broad criteria from Qatar's vision and the institute's mission. The first two criteria determine whether the research topic focuses on Qatar's sustainable development:

- *Does the topic address challenges that a critical industry faces, or does it pave the way for new industries in Qatar related to energy, water, or the environment?* A critical industry is one on which Qatar's economy depends. Addressing challenges associated with the industry includes expanding the market for the industry's products, generating a new revenue stream or industry, making the industry more efficient, and reducing the industry's impact on Qatar's environment and resources. A research topic can also be a priority if it spurs innovation and paves the way for new industries that would help diversify Qatar's economy, which, today, primarily consists of its natural gas and oil industries.
- *Does the topic address challenges that Qatar's critical energy, water, or environmental resources face?* We define a critical resource as one that is abundant in Qatar (e.g., solar energy), is important for development and human health (e.g., groundwater and clean air), drives Qatar's economy (e.g., natural gas), or is a valuable part of Qatar's ecology (e.g., deserts and coastal waters). Addressing chal-

---

[7]   Bartis, 2004.

lenges associated with a resource includes conserving or restoring the resource, improving understanding of the resource, and enabling sustainable use of the resource.

We emphasize *critical* industries and resources to ensure that the institute's work addresses Qatar's most-pressing concerns. Beyond this, these criteria are intentionally general because Qatar's sustainability goals are themselves general. Qatar has not yet measurably specified the balance it seeks to strike between development and conservation (e.g., between coastal development and protection). It has also not determined the *means* by which it would balance development and conservation (e.g., achieving water security by reducing demand, increasing supply, or both). Both of these issues require research, and determining Qatar's research priorities is an important first step in this process.

The next two criteria determine whether the research topics fall within the mission of the institute:

- *Is the topic an applied area of research?* The institute's mission is to conduct applied research that will have an effect on Qatar's critical industries and resources. Therefore, the experts focused on applied research when suggesting potential topics, and this criterion was largely met prior to the topic assessment in step 4.
- *Is Qatar well positioned to pursue this research topic?* The institute should also focus on research in which it could be a regional or global leader. For example, Qatar might have a strategic advantage in certain industries or resources, or Qatar might already have a leadership position in certain subtopics of the research topic. This would provide a competitive research advantage to the institute.[8]

We used these four criteria in steps 4 and 5 to recommend research topics for the institute.

---

[8]   Although they are important, we exclude financial constraints from this criterion because Qatar has expressed a commitment to allocating substantial financial resources to the institute. We also exclude related human-capital constraints. Although this is a significant concern in Qatar and in the region, Qatar has expressed a commitment to acquiring and developing the human capital necessary to achieve its research goals.

Step 4: Assess and Recommend Topics in Separate Panels

We next asked our experts to assess and recommend the research topics using the outputs of background steps 1–3. That is, we used the list of potential topics (from step 1) to seed the assessment and recommendation of topics in the context of Qatar's energy, water, and environment (from step 2) using criteria (from step 3).

We collaborated with subject-matter experts in four discussion panels. The RAND experts participated in three separate energy, water, and environment discussion panels.[9] After the three RAND panels, we facilitated an external panel of experts from CMU to assess energy, water, and environmental issues in a single session.

In each panel, experts reviewed and refined the collection of potential research topics and the organization of those topics. The project team next presented a summary of Qatar's energy, water, and environment goals and challenges and presented the assessment criteria.

The experts then discussed the importance of different research topics in meeting Qatar's sustainability goals and challenges. We encouraged them to reorganize the list of potential topics and to consider topics not on that list. This enabled them to recommend research topics that were tailored to Qatar's unique needs.

Each panel ultimately reached consensus. In total, the panels identified 23 priority research topics that the institute should address before any others. Table 2.3 lists each of these topics, grouped into energy, water, and the environment. Many of these recommended priority topics are a synthesis of multiple items in the list of potential topics in Appendix A, while some are not on that list at all. This reflects the panels' work in developing recommendations tailored to Qatar and to the institute's goals. For each topic, we note in the "Panels' Recommendations" column of Table 2.3 whether there was consensus across panels that the topic should be a priority ("Priority") or whether recommendations were mixed across panels ("Mixed"). In the "Synthesis Outcome" column, we also note the outcome from the project team's

---

9   Individual RAND researchers generally participated in one or two of the three panels on energy, water, or the environment, depending on their areas of expertise.

synthesis of recommendations across panels (step 5). We present details regarding these priority recommendations in Appendix B.

During discussions on priority research topics, experts also offered several insights about secondary research topics—those that could yield benefits to Qatar but that do not address the most-important national challenges. They also noted crosscutting topics—those that are relevant to the other research institutes that QF is considering and thus offer opportunities for cross-institution collaboration. The lists of secondary and crosscutting research are suggestive rather than conclusive. We present these expert recommendations as well in Appendix B.

Table 2.3
Table of Panels' Recommended Priority Topics and Outcome from Synthesis

Topic Area	Topic	Panels' Recommendation	Synthesis Outcome
Energy topics (11)	Natural gas	Priority	Priority (topic 1)
	Petroleum	Priority	Priority (topic 2)
	Carbon capture and storage	Priority	Priority (topic 3)
	Solar energy	Priority	Priority (topic 4)
	Advanced hydrogen and fuel cells	Mixed	Fuel cells are priority (topic 5); hydrogen is secondary
	Green buildings	Priority	Priority (topic 6)
	Smart grids	Priority	Priority (topic 7)
	Strategic energy planning	Priority	Priority (topic 8)
	Energy resource characterization	Priority	Included in strategic energy planning (topic 8)
	Efficient water use in energy production	Priority	Included in environmental characterization (topic 13)
	Polymers, aluminums, and plastics	Mixed	Secondary

Table 2.3—Continued

Topic Area	Topic	Panels' Recommendation	Synthesis Outcome
Water topics (4)	Desalination	Priority	Priority (topic 9)
	Groundwater sustainability	Priority	Priority (topic 10)
	Water demand management	Priority	Priority (topic 11)
	Integrated water resource management	Mixed	Priority (topic 12)
Environment topics (8)	Environmental characterization	Priority	Priority (topic 13)
	Ecosystem services	Mixed	Subtopic in environmental characterization (topic 13)
	Technology procurement	Mixed	Noted in environmental characterization (topic 13)
	Environmental economics	Priority	Subtopic in crosscutting environmental research (topic 14)
	Climate change	Priority	Subtopic in crosscutting environmental research (topic 14)
	Mass–energy balance	Priority	Subtopic in crosscutting environmental research (topic 14)
	Industrial ecology	Priority	Subtopic in crosscutting environmental research (topic 14)
	Closed agriculture systems and advanced aquaculture	Mixed	Secondary

Step 5: Synthesize Findings Across Panels

The project team next synthesized the findings across all four panels to recommend a final set of priority research topics. Table 2.3 summarizes the outcomes of the synthesis process.

We compared the recommendations of individual panels and found that they agreed substantially on the topics that should be a priority for the institute.[10] Nevertheless, there were cases in which their results differed, typically for one of two reasons. First, one panel might have recommended a subtopic (e.g., ocean monitoring) as high priority, whereas another panel recommended the overarching topic as high priority (e.g., environmental characterization). When this occurred, we recommended the broader topic as a priority and the narrow topic as a subtopic.

Second, one panel might have recommended a topic as a priority that other panels did not recommend at all. When this occurred, we reviewed the rationale and the specific language that the panel used in framing its recommendations. Typically, we found that the panel that recommended the research topic used a lower threshold when labeling any research topic as a priority. When we applied a consistent threshold, we found that the recommended research topic was indeed promising but not a priority because it did not address a critical resource or industry. We instead recommend it as a secondary research topic.

For example, one panel noted that the institute *could* conduct research on polymers, aluminum, and plastics as a priority, noting that the abundance of energy and fossil fuel feedstocks *could* enable Qatar to develop industries around this research. It labeled this research as a priority for Qatar. In contrast, this and other panels noted that developing solar energy *should* be a high priority given its abundance in the region. This difference in language suggests that research on polymers, aluminum, and plastics is of secondary importance.

---

[10] Although RAND panels focused on separate areas (energy, water, and the environment), these areas are closely related, and experts in each panel frequently noted priority topics in the other areas. Thus, our comparisons in each area included recommendations across all four panels.

For any other discrepancies, we followed up with the panels to come to a consensus. For instance, one panel recommended fuel cell research as a high priority, while another recommended it as a secondary area of research. We followed up with experts in the latter panel, who, after subsequent review, agreed that fuel cells were a key area of research because of their effect on natural gas.

Finally, we aggregated closely related research topics that would almost always be undertaken together. For example, panels recommended research on environmental economics and developing a mass and energy balance for Qatar. These topics are closely related and inform each other; we therefore recommend them as a combined topic on crosscutting environmental research (topic 14).

Ultimately, we recommend that QF consider 14 priority research topics, which are discussed in Chapters Four through Six. We also note several secondary and crosscutting topics in Chapter Seven. Appendix B describes in greater detail how we synthesized each of the panel's recommended topics into these final recommendations.

Step 6: Report Findings in White Papers

Together, RAND experts and the research team prepared white papers to discuss, in depth, recommendations for each of the 14 priority research topics. Each white paper describes the research topic and provides background information about Qatar that motivates the need for research in the area. Each paper then describes research subtopics that are particularly relevant for Qatar and concludes with a summary of the broader importance of this research in the GCC and the global community. The papers also include opportunities for collaboration and describe the human-capital needs for conducting the recommended research. These white papers are presented as priority-topic subsections in Chapters Four through Six.

CHAPTER THREE

Findings on Environment and Energy Research in the Gulf Cooperation Council

In the first phase of the project, RQPI surveyed Gulf countries and reviewed the literature to identify major research institutions in the GCC member states and to determine the energy and environment research these institutions are undertaking or planning to undertake. Where possible, we assessed missions, the extent to which they collaborated with each other, and other institutional features. We have used these observations to provide guidance for the institute. This chapter presents overall findings and recommendations. We provide profiles of individual institutions in Appendix C.

Gulf Cooperation Council Countries Share Common Resources and Common Concerns

Countries within the GCC have common resources and concerns that are a result of shared geography, culture, and development. Common natural resources among GCC countries include sunlight, water from the Gulf, crude oil, and natural gas. Qatar, Bahrain, the UAE, and the Kingdom of Saudi Arabia (KSA) all have coastlines on the Gulf, giving them shared access to fish stocks and salt water for desalination, while Oman has a long coastline along the Indian Ocean. Sunlight is an abundant resource, presenting opportunities for developing solar energy. GCC countries have benefited significantly from their abundant hydrocarbon resources.

Sustainability is a shared concern among all GCC countries. The desert climate and regional geology limits the natural supply of freshwater and domestic capacity for food production. Most countries in this region must augment their supply of natural water with desalinated water for both human consumption and agriculture. They also share a concern for protecting coastal resources from pollution to sustain fish stocks as a food supply. Additionally, although GCC countries have large hydrocarbon resources, growing populations have sharply increased energy demands from buildings and construction, making renewable energy sources and efficient energy key concerns.

Desert climates, climate change, and expanding populations also raise environmental concerns, such as desertification, which further reduces agricultural output. Environmental health, particularly air quality, is a concern due to rapid industrial expansion and high levels of particulate matter in the air.

Gulf Cooperation Council Institutions Are Engaged in a Range of Research Activities to Address These Concerns

The institutions we reviewed engage in a range of energy, water, and environment research to address these challenging issues.[1] Table 3.1 summarizes the research that is being undertaken in those institutions. Institutions are grouped by country, beginning with Qatar and proceeding alphabetically. We list MEDRC, a multinational institution, last.

As the table shows, there are many common areas of research across countries and institutions. Common energy research includes developing solar energy, improving oil and natural gas recovery and reservoir management, developing carbon capture and storage (CCS), and using energy more efficiently with green buildings and smart grids. Water research largely focuses on desalination, increasing water efficiency, and wastewater management. Environmental research addresses air quality, marine and coastal environments, and conservation and rehabilitation.

---

[1]   We draw out water research as its own category given its importance in the region.

Table 3.1
Gulf Cooperation Council Institutions' Energy, Water, and Environment Research

Location	Institution	Energy	Water	Environment
Qatar	Texas A&M University at Qatar	EOR, extraction technology, fuel cells, GTL, reservoir management, smart grids, solar technology, wind power	Wastewater management	Marine and coastal environment, waste management
	Qatar University	Alternatively fueled vehicles, CCS, energy efficiency, GTL, green building, smart grids, solar energy, waste to energy	Wastewater management	Climate change, waste management
Bahrain	BCSR	—	Water resource planning	Environmental health, food and agriculture, marine and coastal environment
	University of Bahrain	Biofuels, green building, solar technology	Desalination, water resource planning, wastewater management	Conservation and rehabilitation, environmental health, marine and coastal environment
KSA	KACST	Green building	Water distribution, wastewater management	Food and agriculture
	KAUST	Clean combustion, solar technology	Desalination, wastewater management	Marine and coastal environment (planned)

Table 3.1—Continued

Location	Institution	Energy	Water	Environment
KSA, continued	KFUPM	Energy storage, fuel cells, green building, solar technology, wind power	Waste management, efficient water use	Air quality, conservation and rehabilitation, marine and coastal environment
	King Faisal University	—	Desalination, efficient water use, water distribution, water resource planning, water quality, wastewater management	Climate change, waste management
	King Saud University	Alternatively fueled vehicles, CCS, extraction and processing of natural gas and oil, green building, reservoir management, smart grids, sustainable-energy portfolio, wind energy	Desalination, efficient water use, water distribution	Waste management
Kuwait	KISR	Green building, hydro and tidal power, smart grids	Efficient water use, wastewater management, water quality	Air quality, arid-land agriculture, conservation and rehabilitation, environmental health, marine and coastal environment, waste management desalination

Table 3.1—Continued

Location	Institution	Energy	Water	Environment
Kuwait, continued	Kuwait University	Biofuels, CCS, energy efficiency, EOR, green building, reservoir management	Desalination, groundwater sustainability, water quality	Air quality, conservation and rehabilitation, environmental health, marine and coastal environment, waste management
Oman	SQU	EOR, green building, reservoir management	Desalination, efficient water use, groundwater sustainability	Conservation and rehabilitation, environmental health, waste management
UAE	Masdar Institute of Science and Technology	Fuel cells, green building, smart grids, solar technology	Groundwater sustainability, water resource planning, wastewater management	Waste management
	UAEU	Alternatively fueled vehicles, CCS, energy efficiency, energy storage, EOR, green building, reservoir management. smart grids	Desalination, groundwater sustainability, water distribution, water resource planning, water quality, wastewater management	Air quality, conservation and rehabilitation, environmental health, food and agriculture, marine and coastal environment, waste management
International	MEDRC	—	Desalination, water resource planning, efficient water use	—

NOTE: EOR = enhanced oil recovery. BCSR = Bahrain Center for Studies and Research. KACST = King Abdulaziz City of Science and Technology. KAUST = King Abdullah University of Science and Technology. KFUPM = King Fahd University of Petroleum and Minerals. KISR = Kuwait Institute for Scientific Research. QSTP and Center of Excellence for Applied Research and Training (CERT) are not listed because we were unable to obtain information on their research programs.

The research in the region addresses these issues through technology development, policy analysis, and resource management. *Technology development* includes, for example, improving the efficiency and cost-effectiveness of processes for converting natural gas to liquids.[2] Policy research addresses, for instance, the relative merits of incentives and regulations to reduce the demand for water.[3] Resource management includes research to understand and monitor marine life in the Persian Gulf.[4] Research projects can span these categories. For example, projects that evaluate government policies for groundwater use couple public policy and resource management research. Efforts to use new technologies for monitoring oil reservoir capacity combine resource management and technology research.

Activities also range from basic science to applied research to commercialization of technologies. Table 3.2 shows how the research in different institutions varies along this dimension.[5] Institutions are grouped by country, beginning with Qatar and proceeding alphabetically. We list MEDRC, a multinational institution, last.

Much of the work is applied, but many academic institutions are conducting basic science research also. Some institutions, such as SQU, have created centers of excellence to bridge gaps between basic and applied research. There is also interest among university staff in generating patents from applied research through an associated technology-transfer arm.

Many of the institutions we surveyed have a broader mandate than managing a research portfolio. Some are also involved in funding and policymaking. Table 3.2 shows the range of mission areas from the various institutions. QNRF, for example, is purely a funding body, but it has funded a number of applied energy and environment research

---

[2] Texas A&M's project "Development of Novel Gas-to-Liquid Technology in Near-Critical and Supercritical Phase Media" is described at QNRF, 2010.

[3] See, for example, Zekri, 2008.

[4] See, for example, Bishop et al., 2008.

[5] These categories and definitions are adapted from the Organisation for Economic Co-Operation and Development's (OECD's) Frascati Manual (OECD, 2002) for evaluating research and development.

Table 3.2
Gulf Cooperation Council Institutions' Missions in Research, External Funding, and Policymaking

Location	Institute	Environment and Energy Research			External Funding	Policymaking
		Basic	Applied	Commercialization		
Qatar	QNRF				x	
	QSTP		x	x	x	
	Ministry of Environment				x	x
	QU	x	x			
	Texas A&M University at Qatar	x	x			
Bahrain	BCSR	x	x		x	
	University of Bahrain	x	x			
KSA	KACST	x	x	x	x	x
	KAUST	x	x			
	King Faisal University	x	x			
	KFUPM	x	x			
	King Saud University		x			

Table 3.2—Continued

Location	Institute	Environment and Energy Research				
		Basic	Applied	Commercialization	External Funding	Policymaking
Kuwait	KISR	x	x	x		x
	Kuwait University	x	x			
Oman	SQU	x	x			
UAE	Masdar Group and Masdar Institute of Science and Technology	x	x	x	x	
	UAEU	x	x	x		
	CERT		x	x	x	
International	MEDRC		x[a]		x	

[a] MEDRC is currently only funding research, but it has the facilities and plans in place to conduct on-site applied research.

projects. The Ministry of Environment in Qatar develops national environmental policy, sets research priorities, and directly funds some applied environmental research.

Like Others, Qatar's Institute Should Take a Multidisciplinary Approach

Real-world energy and environment challenges are multidisciplinary. Air- and water-quality research might involve information scientists and engineers to develop monitoring systems, physical and environmental scientists to model fate and transport of pollutants, and health scientists to assess effects on humans.

Accordingly, many of the institutions that we contacted take a multidisciplinary approach to research and study through multidisciplinary degree programs or integrated centers of excellence.[6] They are also developing institutional mechanisms to foster multidisciplinary teams. For example, at KISR, researchers in a particular discipline may work in any one of its four research centers.[7] KISR recently initiated a program in which one individual in each discipline serves as a "mentor" and coordinates activities of all researchers in that discipline across all the centers. This approach promotes cohesiveness within a particular discipline so that colleagues share and disseminate knowledge, while also promoting multidisciplinary teams within centers.

We recommend that the institute similarly take an intentionally multidisciplinary approach to its research and focus its work around problem statements or topical centers of excellence.[8] Additionally, there might be opportunities for the institute to collaborate with QF's two other research centers: the Qatar Biomedical Research Institute and

[6] For instance, part of the Masdar Institute of Science and Technology's mission statement is to "[e]stablish and continually evolve interdisciplinary, collaborative research and development capability in advanced energy and sustainability" (Masdar Institute, undated [a]).

[7] The four research centers at KISR are Water, Petroleum, Energy and Buildings, and Environment and Life Science.

[8] Researchers may still be affiliated with academic or disciplinary departments, as in a matrix organization.

the Qatar Computing Research Institute. In Chapter Seven, we suggest topics for which such cross-institutional collaboration might be appropriate.

Common Research Interests and a Shortage of Human Capital Suggest That the Institute Should Collaborate

The common research interests among GCC institutions lead to some duplication of research. Duplication in critical areas of research might be desirable and lead to faster, better, or more-diverse solutions to energy and environment problems. It might also help to build local human capital, which is a current concern and a long-term goal for many GCC countries.

Indeed, perhaps the strongest message from our GCC research was that shortages of human capital are the biggest hurdle to achieving regional research objectives: Although institutions, facilities, and funding are in place, it is difficult to recruit qualified researchers to staff them. Recruiting is competitive not only within the GCC but also with Europe and the United States.

Common research interests and a shortage of human capital suggest a great need for collaboration and coordination among GCC institutions and with institutions farther abroad, which could also increase the capacity for quality research. For example, the region could become the leader in solar energy research, development, and testing. A critical mass of collaborating researchers from different institutions in the region in such a key research area could help attract other top-tier scientists from farther abroad to the region. A coordinated research portfolio could further enable different institutions to address a shared energy or environment challenge from different but complementary directions.

All interviewees and most of the publicly available information we reviewed suggested a desire for greater collaboration.[9] However, actual collaborative efforts between institutions within the GCC and even within the same country have been limited.[10] This is in part due to bureaucratic requirements and limits on uses of funds. We found that institutions in the GCC were more likely to have partnerships with universities and other institutions beyond the GCC than they were to have partnerships with each other. Some of these partnerships are based on historical relationships with the hydrocarbon industry, as is the case with KISR's partnership with Japan. Other relationships have been built to take advantage of established capabilities in Western institutions (for instance, the Masdar partnership with the Massachusetts Institute of Technology [MIT]).

We recommend that the Qatar Environment and Energy Institute develop relationships with institutions and other scientists in the region. This could be achieved through formal memoranda of understanding (MOUs) between institutions or more informally through regional conferences, workshops, or forums. We suggest that QF draw from the GCC Energy and Environment Research Database we have developed to identify potential institutional partners.

We also urge QF to develop the institute's research portfolio and programs in close coordination with other institutions in the country and collaborate when possible. This will help avoid duplicate research where it is not needed. This is a particular concern in key areas in which Qatar is already heavily engaged, such as natural gas research. Sharing human capital can help avoid competition between institutions, which can be counterproductive to addressing research problems. Along with shared facilities, this can also help the institute undertake research quickly.

---

[9] For example, according to Kuwait University's *Research Administration: Annual Report 2008–09*, one of the university's key goals is "to strengthen its scientific horizons through collaborative linkages with distinguished institutions and experts, and to perpetuate the culture of enhancing and enriching institutional capabilities through effective alliances and partnerships" (Kuwait University, 2009, p. 53).

[10] We did find a formally established partnership between SQU in Oman and UAEU in Al Ain.

The Institute Should Seek Input from and Involve the Qatari Government

Many of the most-pressing energy, water, and environmental challenges are issues of public well-being or public good, such as air and water quality or biodiversity. Research aimed at developing solutions to pressing problems in these areas often lack clear markets, making them hard to commercialize and less likely to be taken up by industry.

Thus, government funding and policymaking are needed to promote research that could have long-term benefits and promote the public good, particularly for environmental protection, pollution monitoring, and in the development of approaches to manage national resources, such as water.

Additionally, government is a key stakeholder and a potential collaborator for research in these areas, particularly for public-policy research. For example, research on water demand–management policies might depend on government data about water consumption. In turn, the government might use findings on the acceptability and effectiveness of different policy options to frame its own policies. Researchers and government officials might work together to implement those policies. Thus, it is important that research institutions and governments align their research priorities and collaborate. Additionally, we found in our survey that government funding priorities might not be clearly identified or well integrated with existing research interests or capabilities.[11]

We recommend that QF seek input from and involve the Qatari government. The Qatari government recently articulated some of its energy and environment priorities in the launch of its *National Development Strategy* for 2011–2016.[12] The institute's overall research portfolio should be consistent with these and other priorities. Addition-

---

[11] For example, researchers at one university stated concerns that the government had not provided clear direction on priority areas and that much government funding went to researchers' "pet projects" that were not necessarily in the best interest of the country.

[12] We developed our recommendations for the institute's research portfolio in parallel with the recommendations made by the National Development Strategy Committee. The *National Development Strategy* was launched in March 2011.

ally, the institute should engage with government agencies, such as the Ministry of Environment and the Ministry of Energy and Industry, and others to ensure that

- research priorities are aligned in specific areas
- resources allocated to energy and environment research reflect the institute's capabilities and needs
- research findings reach their intended audience
- the institute and government agencies collaborate.

CHAPTER FOUR

Priority Energy Research

Overview of All Priority Research Topics

We have identified 14 priority research topics that the Qatar Foundation should consider for the Environment and Energy Institute (Table 4.1). These topics address Qatar's most-pressing issues in energy production and use, water production and use, and environmental preservation. These topics should be thought of as a menu of options from which QF should choose. That is, although experts consider the recommended topics to be priorities for Qatar, we are not recommending that the institute pursue all of them, at least not initially. Rather, these topics should be used as a basis for QF to develop an initial research program that can grow over time. To develop the initial program, the Qatar Foundation should consider additional criteria, such as the available human capital, costs and benefits of different research portfolios both now and in the future, near-term collaboration opportunities, research being considered elsewhere (to avoid unnecessary duplication), and Qatar's evolving sustainable development priorities. These issues are discussed in Chapter Eight.

We discuss these research topics in three chapters. In this chapter, we discuss energy research; in Chapter Five, we discuss water research; and, in Chapter Six, we discuss environmental research. For each topic, we describe the importance of the research topic in addressing Qatar's opportunities and challenges, note key research subtopics, and summarize potential leadership and collaboration opportunities.

It is important to note that, although we have organized these topics in separate categories, they are interrelated. Many topics address

Table 4.1
Priority Research Topics for Qatar's Environment and Energy Institute

Research Area	Topic
Energy (Chapter Four)	1. Natural gas production and processing
	2. Petroleum production and processing
	3. CCS
	4. Solar energy development
	5. Fuel cells
	6. Green buildings
	7. Smart grids
	8. Strategic energy planning
Water (Chapter Five)	9. Desalination
	10. Groundwater sustainability
	11. Water demand management
	12. IWRM
Environment (Chapter Six)	13. Environmental characterization
	14. Crosscutting environmental research

NOTE: Numbers correspond to topic numbers in this book and are not intended to convey any sense of priority among the topics.

the same concerns but in different ways; for example, research in green buildings (topic 6) and smart grids (topic 7) is aimed at increasing energy efficiency through different mechanisms. Other topics have cascading effects. For instance, sustainable management of Qatar's groundwater (topic 10) could require using desalination (topic 9) for agriculture, which, in turn, could exacerbate the problem of brine, harming coastal ecosystems (topic 13). We have highlighted many of these interrelationships in individual topic discussions.

Overview of Priority Energy Research Topics

Oil and natural gas are Qatar's critical energy resources, and they are the backbone of Qatar's economy and its critical industries. With proven oil reserves of more than 15 billion barrels and natural gas reserves of nearly 900 trillion cubic feet, Qatar is one of the world's leading exporters of these fossil fuels.[1] Oil and natural gas have facilitated Qatar's rapid development in recent decades, and Qatar has used revenues from those sources to develop its infrastructure and human capital.

Today, Qatar faces important opportunities and challenges in oil and natural gas. First, Qatar's oil and gas are limited resources, and Qatar is seeking ways of using these resources efficiently and to the greatest benefit for its economy and population.[2] For instance, Qatar exports much of its natural gas as liquefied natural gas (LNG). However, there are significant logistical barriers to exporting LNG, suggesting that alternative natural gas products might be more-viable exports. Qatar also uses natural gas to produce electricity domestically. This is the most cost-effective way of producing electricity in Qatar, but the opportunity cost of export losses could be significant. A key question that could be addressed through research is how Qatar should balance exploitation and conservation of its fossil fuel resources, the production of competing fossil fuel products, and domestic use versus exports of fossil fuels, now and in the future.

Second, fossil fuels are responsible for approximately 75 percent of global anthropogenic (human-caused) GHG emissions from the past 20 years.[3] Thus, although fossil fuels remain the primary source of energy globally, demands for clean energy—including clean fossil fuels—are growing. Natural gas burns more cleanly and emits fewer GHGs than other fossil fuels, suggesting that natural gas could be an ever more valuable resource in a world in which carbon emissions are increasingly regulated. On the other hand, petroleum—which has high

[1]   Energy Information Administration, 2011a, 2010.

[2]   General Secretariat for Development Planning, undated (b).

[3]   Energy Information Administration, 2004.

carbon emissions—will be disadvantaged by such policies. Reducing emissions from the petroleum life cycle could be key to gaining a competitive advantage in the petroleum market.

Third, as Qatar's fossil fuel reserves are depleted, Qatar will need alternative, renewable forms of energy. Solar energy is the only significant renewable energy resource in Qatar, but it is plentiful there and in the region more generally. If solar energy can be substituted for fossil fuels or added to fossil fuel processes to make them more efficient, significant GHG emissions could be avoided. Qatar would have a reliable source of clean, renewable energy well into the future. However, many kinds of solar energy conversion are presently too expensive and are technologically unable to realize this full potential, while others have not been sufficiently deployed. Much research and development is needed to harness the potential of this resource for the country.

We recommend that the institute help Qatar address these challenges and capitalize on opportunities by conducting research in several key areas. The institute should improve the production and processing of natural gas (topic 1) and petroleum (topic 2) and help Qatar use CCS (topic 3) to further improve these fossil fuels. It should develop solar energy (topic 4) to offset the use of these resources, and it should develop fuel cells to improve electricity generation in the longer term (topic 5). The institute should conduct research that increases domestic energy efficiency (e.g., through the use of green buildings [topic 6] and smart grids [topic 7]). Finally, the institute should help Qatar develop a long-term strategic energy plan (topic 8).

Priority Topic 1: Natural Gas Production and Processing Research

Background and Motivation

As of January 1999, Qatar had more than 25 trillion cubic meters (m^3) of proven natural gas reserves, the third largest in the world behind Russia and Iran.[4] Qatar has developed its natural gas reserves signif-

---

[4] Energy Information Administration, 2011a.

icantly over the past two decades. In 1995, Qatar produced 14 billion m³ of natural gas, and its exports of natural gas were minimal. In 2009, Qatar produced 80 billion m³ of natural gas and exported more than 68 billion m³ of that, 49 billion m³ (73 percent) as LNG.[5] Revenues from natural gas exports have increased similarly. Total current capacity for LNG production in Qatar is 76 billion m³ per year.[6] In comparison, petroleum exports have roughly doubled over the same time period.

Qatar has used revenues from natural gas to develop its infrastructure and human capital. Among these investments are research programs, such as the institute itself. Because natural gas plays a critical role in supporting and shaping Qatari society, research in natural gas should have the strategic goal of enhancing the inherent advantages of natural gas and overcoming challenges.

Natural gas has several advantages over other fossil fuels. Natural gas burns more cleanly and emits fewer GHGs than other fossil fuels do. For example, when used to produce electricity, natural gas emits significantly fewer pollutants and approximately 50 percent less GHG than coal. Natural gas–fired power plants are less expensive to construct and operate than coal-fired power plants because of the absence of pollution-control and material-handling equipment. Because of this, natural gas is already a preferred fuel for electricity generation in the United States and elsewhere. And a carbon-constrained world in which emissions of GHGs are regulated would further favor natural gas.

Simultaneously, there are resource, logistical, economic, and environmental challenges to the increased development of natural gas in Qatar:

- Resource challenge: ensuring long-term supplies of natural gas. Though Qatar is endowed with significant resources, its most pressing strategic priority is to manage this resource well.
- Logistical challenge: bringing natural gas to market. Natural gas is typically transported by pipeline. However, Qatar's location

---

[5] British Petroleum, 2010.

[6] Energy Information Administration, 2011a.

is far from demand centers, and pipeline capacity from Qatar is limited. Absent another means of transport, this gas is known as *stranded*. Qatar addressed this challenge by building capacity for liquefying natural gas and designing and building a fleet of pressurized and insulated tankers to deliver the natural gas. Nations that import LNG, and particularly nations that are considering importing LNG in the future, are concerned about the safety and security of LNG-receiving facilities. Synthetic fuels produced from natural gas can be transported by a standard tanker, alleviating transportation issues, and can meet demands for refined-petroleum products, expanding the market for Qatari gas.

- Economic challenge: developing high-value natural gas products. In addition to its logistical challenges, conventional natural gas has a lower price on the global market than petroleum or refined-petroleum products.[7] This means that Qatar receives less revenue from conventional natural gas than it does from petroleum, despite significant increases in natural gas production in the past two decades. Producing synthetic fuels and other petroleum products from natural gas can therefore be more profitable than exporting LNG.

- Environmental challenge: Reducing pollution and mitigating GHG emissions. Production, processing, and use of natural gas results in air and water pollution. Natural gas is significantly cleaner than other fossil fuels, and pollution-control technologies are commercially available. Managing emissions of GHGs, however, is essential to maintaining natural gas's long-term competitiveness.

These challenges could limit the expansion of Qatari natural gas and products. From what we know about current and expected development of natural gas in Qatar, we suggest that the institute focus on

---

[7]   September 2010 global price for Qatari Dukhan crude was $76 per barrel, or approximately $13 per million British thermal units (mmBtu). In contrast, the price of imported LNG in the United States was $4.5 per mmBtu over the same time period. These prices include transportation, so the price received in Qatar for either petroleum or LNG would have been less.

two strategic issues that will help advance the status of Qatar's natural gas compared with that of other fossil fuels and address the key challenges:

- First, although the life-cycle GHG emissions of natural gas are naturally lower than those of other fossil fuels, in a carbon-constrained world, the price of all fossil fuels is likely to rise. The institute should undertake research to minimize GHG emissions of natural gas and products derived from natural gas. Such research would further improve natural gas's competitive position with respect to other fossil fuels and some kinds of renewables, thereby helping to address economic challenges. This research would also help address the environmental challenge of climate change.
- Second, the institute should conduct research to enable Qatari natural gas to meet growing demands for specialized refined and synthetic products and on improving LNG operations. Such research would help to address logistical, environmental, and economic challenges. For example, synthetic fuels are compatible with existing infrastructure, handling systems, and vehicles, making them logistically easier to export than LNG is. They also have comparable life-cycle GHG emissions.[8] And, synthetic fuels have a higher market value than natural gas does.

We discuss each of these research areas in greater detail in the next sections. Lastly, to address resource challenges, we recommend that the institute work directly with Qatar Petroleum and Qatargas to develop and undertake complementary research programs.

Recommended Research: Managing Greenhouse Gas Emissions from Natural Gas

The institute should undertake research on reducing upstream and downstream GHG emissions from natural gas. As we have mentioned, this includes improving efficiency of natural gas production, improving

---

[8] Hileman et al., 2009.

operations to prevent fugitive emissions, enabling CCS, and improving end-use technologies.

Improving Efficiency of Natural Gas Production and Processing. The institute should undertake research to improve existing, and develop new, technologies that reduce the amount of energy required to extract and process natural gas. Such research would focus on developing and deploying advanced drilling and recovery technologies. Qatar Petroleum and its partners are already undertaking such research and would be key partners for the institute.

Improving Operations to Prevent Fugitive Emissions. Methane, the principal component of natural gas, has a global-warming potential 25 times that of carbon dioxide (CO_2).[9] For this reason, releases of natural gas from wells and leaks from pipelines need to be managed. There are many sources of fugitive emissions: Flaring and venting certain streams of natural gas might be a safety measure at a drill site or processing facility; leaks can occur at seals and valves at processing facilities and pipelines; leaks can also result from exploration and drilling activities. Managing so-called fugitive emissions is very challenging because of the scale and complexity of natural gas processing and transportation systems. Currently, fugitive emissions are largely estimated by applying gross-emission factors to total production or operations.[10] The emission-factor fugitive emissions of natural gas from sour gas–processing facilities is 9.7 grams of methane per cubic meter of raw gas feed, or 1.4 percent of natural gas at standard conditions. Research in this area includes the development of monitoring and management systems to identify fugitive releases and assist in reducing them, thus reducing life-cycle GHG emissions of natural gas operations. This research would build on existing activities of Qatar Petroleum. Additional applications of such research can be found in natural gas distribution and management, including monitoring systems to detect leaks.

Carbon Capture and Storage. CO_2 is coproduced with natural gas. It is also a by-product of processing natural gas, upgrading coproduced liquids, producing synthetic fuels, and combustion. Many of

---

9   Forster et al., 2007, pp. 130–234.

10   Carras et al., 2006.

these streams of CO_2 can be isolated and purified. They can then be used in EOR. In this process, CO_2 is injected into a field at pressure, where it mixes with the remaining oil in the field, changing its flow properties. The oil can then be pumped from a producing well.[11] The injected CO_2 can be recovered and reused. After completing EOR operations, it is possible to pump CO_2 into the reservoir and close the site, permanently storing the CO_2 and thereby preventing it from contributing to emitted GHGs.[12] CCS is covered in the discussion of topic 4.

Improving Efficiency and Reducing the Environmental Impact of Systems Using Natural Gas. Another area of research is to improve systems that use natural gas, particularly natural gas power plants.[13] Improving component processes of existing cycles and advancing high-efficiency combustion cycles can improve the efficiency of thermal power plants, thus conserving natural gas resources and reducing GHG emissions.[14] Water is also a concern in electricity production. Traditional thermal power plants use significant amounts of water primarily for cooling, for boilers, and for reducing the amount of air pollutants produced.[15] Most power plants in the region use Gulf

---

[11] Toman et al., 2008.

[12] The institute could also consider research into using CO_2 for enhanced natural gas recovery, a process that is not as mature as EOR. In this process, compressed CO_2 would be injected into a nearly depleted natural gas reservoir, repressurizing it and forcing out any remaining natural gas. As in EOR operations, the CO_2 could be recovered or reused. Because natural gas reservoirs are capable of storing gases over geologic timescales, CO_2 can also be stored in depleted natural gas reservoirs.

[13] Natural gas is also used in end-use systems, such as natural gas–fired refrigerators. These are addressed as part of the discussion of green buildings.

[14] Organic Rankine cycle engines use waste heat to produce electricity and could offer another way of increasing the efficiency with which natural gas combustion is used to produce electricity. Organic Rankine cycle engines use a low-boiling-point organic solvent, rather than water and steam, and are therefore typically applicable to lower-temperature thermal resources, such as geothermal or solar thermal turbines. However, they can also be designed to use the lower-temperature waste heat from higher-temperature thermal technologies, including natural gas combustion and stationary fuel cells.

[15] For natural gas plants, this involves spraying water through the combustion process, which lowers the temperature and reduced the nitrogen oxide (NOx) pollution (EPRI, 2002a).

water for cooling,[16] so much of the concern about water used in electricity production in the region is associated with its impact on Gulf ecosystems. We address this further in the discussion of topic 13 on environmental characterization. Nevertheless, freshwater is also used in electricity production (e.g., in steam turbines). The institute should therefore also conduct research to reduce the freshwater used in electricity production.

In the longer term, such technologies as fuel cells might use natural gas and water more efficiently to produce electricity. Fuel cells are addressed in the discussion of topic 5.

Table 4.2 summarizes how the research categories apply to various aspects of natural gas production and use in Qatar.

Recommended Research: Improving Methods of Producing Natural Gas Products

Methods of processing natural gas, refining condensate, producing synthetic fuels, and liquefying natural gas are mature.[17] New research should focus on improving these methods to make them more robust, environmentally friendly, and cost-competitive.

Consider this example: When producing liquid fuels from natural gas, the natural gas is first converted to a synthesis gas comprised of hydrogen and carbon monoxide. The synthesis gas is then passed over a catalyst to produce hydrocarbons. Current industrial methods convert natural gas to synthesis gas by combining methane with oxygen in a process called partial oxidation. This process is very efficient, but it requires a source of pure oxygen. Significant energy is required to produce pure oxygen, and the process is highly exothermic, posing a heat-management and recovery challenge.

---

[16] Parker and Qasim, 2006.

[17] Condensates are natural gas liquids (e.g., ethane, propane, butane) that are coproduced with the natural gas. Qatargas cleans and separates the condensates and provides them as a feedstock to the petrochemical industry.

Table 4.2
Matrix of Research Areas for Reducing Life-Cycle Greenhouse Gas Emissions of Natural Gas Processing and Use

Research Area	Potential Application Area
Improving efficiency of natural gas production and processing	Advanced drilling technologies; improved separation and processing; improved catalysts for processing and synthetic-fuel production
Improving operations to prevent fugitive emissions	Well, pipeline, and distribution-system monitoring
CCS	Assessment of opportunities, costs, and benefits for EOR (see CCS discussion)
Improving efficiency of systems using natural gas	Improved natural gas power cycles; fuel cells (see "Fuel Cells" discussion, topic 5); improved consumer end-use technologies (see "Green Buildings" discussion, topic 6).

Alternative approaches include steam reforming the natural gas and CO_2 reforming.[18] In steam reforming, natural gas and water vapor are reacted in the presence of the catalyst to produce synthesis gas; in CO_2 reforming, methane and CO_2 are combined, but the process is endothermic—that is, it consumes energy. Additionally, it is possible to combine CO_2 reforming and the partial-oxidation approaches, improving the thermodynamics of the process.[19]

Other performance-improvement research could include improving gas cleanup, coproduct separation, and refining. Improving gasification, gas cleanup, coproduct separation, and other processes often requires the development, fabrication, and testing of appropriate catalysts to drive the chemical reactions. Moreover, many catalysts are expensive, are susceptible to poisoning (i.e., damage from reacting with other elements and compounds), and degrade relatively quickly. The institute should develop catalysts that are resistant to poisoning, have longevity, are comprised of less expensive elements, and retain effec-

---

[18] For more on steam reforming, see Gesser, Hunter, and Prakash, 1985. For more on CO_2 reforming, see Ashcroft et al., 1991.

[19] He et al., 2009.

tiveness. Such research would have applications throughout Qatar's natural gas and petroleum industries.

A portion of institute-supported research in the area of process advancement should focus on LNG because it will remain the primary form of natural gas exports. Research in this area could seek to improve efficiency and safety and reduce risks in operations.

Collaboration Opportunities and Human-Capital Needs for Qatar

The institute's research should complement other natural gas research in Qatar. For example, Qatar University is conducting a wide range of natural gas research as part of its Gas Processing Center. This includes materials engineering, natural gas processing, and control and instrumentation. Research at the Texas A&M University at Qatar is largely focused on natural gas extraction and synthesis, including reservoir engineering and managing Qatar's predominant sour (sulfur-rich) gas, reactors, and catalysts. We recommend that the institute consult with these research institutions, as well as with Qatar Petroleum and Qatargas, before undertaking natural gas research, in order to ensure that the institute does not duplicate research being done elsewhere in Qatar.

There are also opportunities to collaborate with institutions in the GCC countries—for example, with KISR in natural gas power-plant research. King Saud University in Saudi Arabia and UAEU are doing a broad range of natural gas research, including liquefied petroleum gas, fuel cells, and efficient natural extraction.

The institute will require expertise in multiple disciplines to undertake this research. Among these are geology and geophysics, physical and catalytic chemistry, materials science, process engineering, operations research, and economics. Qatar has significant natural gas expertise in both the private and public sectors. Qatar University and Texas A&M University at Qatar offer bachelor of science (BSc) degrees in a range of science and engineering disciplines. Universities in the region also offer advanced degrees. For example, King Fahd University for Petroleum and Minerals in KSA offers civil engineering Ph.D.'s with a focus in geotechnical engineering. King Saud University also offers master's-level courses in chemical engineering, industrial engineering, and petroleum and natural gas engineering.

Priority Topic 2: Petroleum Production and Processing Research

Background and Motivation

In 2009, Qatar exported slightly more than 1 million barrels of oil and refined-petroleum products per day.[20] Although Qatar is one of the smaller exporters in the Organization of the Petroleum Exporting Countries, these exports currently generate significant revenue for Qatar (and, until recently, revenue from oil exports exceeded revenues from natural gas exports). This is in part because Qatar's oil is light and highly valued in international markets. Qatar has also been investing in domestic refining capacity to enable it to export higher volumes of refined-fuel products.[21]

Although the expanding natural gas sector has overtaken petroleum as the major segment of the economy,[22] Qatar's petroleum resources will remain important. Qatar currently has approximately 25 billion barrels of proven petroleum reserves.[23] At current production rates and without additions, these reserves represent a 52-year supply of petroleum.

Qatar Petroleum and its international industry partners spearhead petroleum research in Qatar. The research needs and agendas are not public but include improving oil and gas operations, environment and climate change, and reservoir engineering and management.[24] Of these, we believe that reservoir engineering and management, which includes estimating reserve sizes and developing extraction strategies for high recovery rates, is of critical importance to Qatar because its reserves are small relative to its production rates.

The institute should directly and closely collaborate with Qatar Petroleum to identify a complementary research agenda and to execute its research. The institute can play an important role in contributing to

---

[20] British Petroleum, 2010.

[21] Energy Information Administration, 2011a.

[22] "Qatar GDP Statistics Show Gas Revenue Surpassed Oil in 2009," 2010.

[23] Organization of the Petroleum Exporting Countries, 2010.

[24] QSTP, 2008.

long-term, environmentally conscious development and use of Qatar's petroleum resources.

One area of particular promise is EOR by CO_2 flooding, which is also addressed in the discussion of natural gas (topic 1) and CCS research (topic 3). In EOR, high-pressure CO_2 is injected into an oil field, where it mixes with the remaining oil and improves its flow and the ease of extraction.[25] In current applications, it is common to recover the injected CO_2, reinject it, or transport it to another site for use. This is because obtaining sufficient quantities of CO_2 is currently costly in areas employing EOR. However, in Qatar, CO_2 is naturally copro-duced during oil and gas operations and can be isolated relatively easily from synthetic-fuel facilities.

Collaboration Opportunities and Human-Capital Needs for Qatar

There are other opportunities for collaboration on petroleum research in Qatar. For example, Texas A&M University at Qatar is a leader in petroleum research in Qatar. Several ongoing projects concern EOR by CO_2 flooding as a means of enhancing extraction efficiency.

Many research institutions in the GCC countries were founded on the mission of petroleum research. Thus, they might be good can-didates for collaboration with the institute. In KSA, KACST has the Petroleum and Petrochemicals Research Institute, and KFUPM hosts the Center of Research Excellence in Petroleum Refining and Petro-chemicals. KISR also hosts a dedicated Petroleum Research and Stud-ies Center. From the research portfolios we reviewed, we found a sub-stantial focus on EOR in Qatar and the GCC. SQU in Oman and Kuwait University have ongoing projects in the area of EOR. Reser-voir modeling and management research is also being conducted at KFUPM and Kuwait University. On the environmental side of petro-leum research, both Kuwait University and SQU are exploring oil-spill effects and rehabilitation options for hydrocarbon-contaminated groundwater and soil.

Petroleum research requires expertise in the fields of oil and gas engineering and exploration, physical and catalytic chemistry, geol-

---

[25] Toman et al., 2008.

ogy and geophysics, materials science, and mechanical and chemical engineering. Private companies in Qatar, including Qatar Petroleum, Shell, Chevron, and ConocoPhillips, have both research and operational expertise, although most of the international petroleum and petrochemical companies conduct their major research and development activities outside Qatar. Although none of Qatar's universities offers an advanced degree in these areas, there are opportunities to recruit or develop research talent at the postgraduate level throughout the GCC. KFUPM, for example, offers Ph.D. programs in petroleum and chemical engineering. UAEU also offers Ph.D. programs in petroleum and chemical engineering, as well as in materials science. Other related advanced-degree programs can be found at SQU and Kuwait University.

Priority Topic 3: Carbon Capture and Storage

Background and Motivation

Increasing concentrations of GHGs in the earth's atmosphere are driving climate change. GHG emissions consist primarily of CO_2 but also include methane and nitrous oxide. Fossil fuel combustion and deforestation are the primary sources of anthropogenic (human-caused) GHGs. Fossil fuels in particular are responsible for approximately 75 percent of global anthropogenic emissions from the past 20 years.[26]

Fossil fuels remain the primary source of energy globally, and natural gas and petroleum are the backbone of Qatar's economy. However, demands for clean energy—including clean fossil fuels—are growing. Although natural gas has a lower GHG intensity than other fossil fuels, such as petroleum and coal (i.e., it produces less CO_2 per unit of energy), emissions are still significant. Moreover, in a carbon-constrained world in which emissions are regulated, all fossil fuels, including natural gas, will be affected.

CCS (sometimes also called *carbon capture and sequestration*) is the process of capturing CO_2 and depositing it in a reservoir for per-

---

[26] Energy Information Administration, 2004.

manent storage, thereby preventing it from contributing to increases in atmospheric concentrations of GHG and the resulting climate change. CCS is aimed at capturing the CO_2 that is released by fossil fuel combustion and will be the key to making fossil fuel use acceptable in a carbon-constrained world.[27]

Generally, CCS is envisioned as being used in conjunction with stationary applications, such as the production of electricity or making liquid fuels from natural gas (so-called GTL processes).[28] CCS is likely to be able to capture about 90 percent of CO_2 emissions produced in natural gas combustion for electricity production.[29] If Qatar could use CCS to lower the carbon footprint of its own natural gas use, it would reduce Qatar's own CO_2 emissions at home, as well as potentially expand markets for natural gas around the world.

Although CCS technologies are already being actively investigated in Qatar and elsewhere, the institute should undertake two aspects of CCS research that are most relevant to Qatar. First, Qatar must be able to store CO_2. The institute should help determine the geological feasibility and magnitude of storage potential for CCS in Qatar and research the policies and infrastructure necessary to apply CCS in Qatar.[30] Second, the institute should research and develop natural gas CCS technology. These two issues are not likely to be sufficiently

---

[27] There are also GHG emissions associated with extraction and processing of fossil fuels. However, in general, production emissions make up a small fraction of total life-cycle emissions for conventional fossil fuel. Fuel use (i.e., actually burning the natural gas, petroleum, or coal) is much more significant. Nevertheless, in a highly carbon-constrained world, these upstream emissions could also be captured. In some cases, CCS could also be used to reduce production GHGs, but, in this discussion, we assume that the largest source of emissions—combustion—will be dealt with first, in the near term.

[28] It would not be readily applicable to reducing mobile emissions from transportation.

[29] Metz et al., 2005.

[30] The use of CO_2 for EOR is another practice that can offer some early, economically attractive storage for CO_2 and is a well-established practice. In EOR, CO_2 is injected at high pressure into an oil field. There, it mixes with the remaining oil and improves the flow and ease of oil extraction. The CO_2 can then be recovered, reinjected, or transported to another site for use, but a significant fraction of the CO_2 can be left behind in the deposit; depending on site-specific conditions, it can be permanently stored in this way (Toman et al., 2008).

addressed in the near term by research being conducted elsewhere in the world.[31]

Recommended Research: Technical, Logistical, and Policy Issues for Carbon Dioxide Storage in Qatar

For CCS to be viable in Qatar, there must be a way to safely and permanently store CO_2 in the country or elsewhere in the GCC region. There must also be a sufficient physical and policy infrastructure in place before CCS can begin. The institute should begin by identifying and characterizing potential geological storage sites within its borders, such as depleted oil and natural gas reserves.[32] An accurate estimate of the total capacity of all geologic storage sites in Qatar is necessary to assess the potential for CCS.[33]

The institute should also develop techniques for monitoring and verifying (M&V) the short- and long-term stability of CO_2 stored in each site. Although M&V practices are being researched and developed in other places in the world, it is likely that these methods will need to be tailored to the specific geology in Qatar.

Oversight of CCS practices will also be needed—e.g., to ensure that proper safety procedures are being followed. Qatar will likely need to develop official policies and establish an agency or other entity within an existing agency to govern CCS activities. Other logistical and liability concerns, such as identifying responsible parties for the CO_2 at various stages of the CCS process and contingency planning in

---

[31] In the longer term, CCS research at the institute could also include carbon capture, *use*, and storage (CCUS), in which CO_2 is also utilized for other purposes. For example, CO_2 captured from stationary power plants could be used to grow marine algae for biofuels, though a cleaner flue-gas stream is likely to be needed for this purpose than for sequestration. If the institute undertakes marine-algae biofuel research in the future, which we suggest as a secondary topic, then CO_2 use could be a joint area of research with its CCS research program.

[32] Storage of CO_2 can occur in underground deep geologic formations, deep in the ocean, or as a solid mineral carbonate. Geologic storage is likely to offer the best combination of cost-effectiveness and environmental acceptability in the near term. In addition to oil and gas formations, CO_2 can be stored in deep underground saline formations.

[33] If there is not sufficient storage capacity in Qatar, a regional assessment and regional cooperation will be necessary to undertake CCS.

the case of CO_2 leakage, will also need to be addressed.[34] The institute should undertake policy research in these areas, such as assessing best practices for CCS policy and oversight.

The institute should further determine what kinds of physical infrastructure are necessary for CO_2 storage. For example, natural gas power plants could be sited to allow direct access to CO_2 storage sites, but this might require the construction of transmission lines to enable electricity to reach demand centers. It is therefore likely that CO_2 pipelines will be necessary to transport a significant amount of CO_2 to appropriate sites.[35]

Recommended Research: Advancing Natural Gas Carbon Capture and Storage Technology

Because the combustion phase of conventional natural gas use generally dominates the total life-cycle emissions associated with natural gas use, CCS is mostly likely to be applied, at least initially, to the combustion of natural gas for electricity production.[36] Natural gas CCS technology could help Qatar achieve long-term, ultralow-GHG energy production. However, most global research focuses on integrating CCS with direct coal combustion (or, to a lesser extent, coal gasification) for two reasons. First, coal-based power dominates in many other parts of the world, such as in the United States and China. Second, natural gas already has a relatively low carbon intensity compared with other fossil fuels.[37] Therefore, the institute's research niche might be in developing CCS technology that is optimized for natural gas.

One approach to combining natural gas with CCS is postcombustion capture. In postcombustion capture, natural gas is first burned to generate steam and produce electricity. Then, the flue gas—i.e., the postcombustion exhaust that contains CO_2 and other wastes—

---

[34] Metz et al., 2005.

[35] Metz et al., 2005.

[36] As noted, natural gas extraction and processing emissions are generally relatively small, and CO_2 produced in transportation use of fossil fuels is not amenable to CCS.

[37] Specifically, natural gas emits about 50 percent less carbon per unit of electricity delivered than coal does (Metz et al., 2005).

undergoes CO_2 separation to extract a relatively pure stream of CO_2 that can then be compressed and stored. Postcombustion CCS can be integrated into an existing CO_2-producing power plant as a retrofit, or it can be incorporated into new construction. For retrofits, the design depends on the efficiency of natural gas combustion technology in place. For existing turbines that are highly efficient and relatively new, it is possible that the benefit of adding postcombustion CCS would not be worth the cost until a significant premium were placed on CO_2 reduction (approaching $100 per ton of CO_2 avoided). For new natural gas–fired electricity plants, a natural gas combined-cycle (NGCC) turbine with CCS is promising.[38] Such a system has been considered but to date has not been deployed. This design is estimated to have slightly lower costs than retrofitting an existing NGCC plant, at around $80 per ton of CO_2 avoided.[39]

The institute should research ways to optimize these postcombustion designs, pilot-test promising concepts, and develop ways of reducing the costs of these technologies. Because these are emerging technologies, there is likely to be opportunity for making significant contributions—for example, by increasing the fundamental efficiency of the CO_2 separation processes—or for building first-of-a-kind facilities.

The institute should also undertake research in CO_2 separation technologies, especially if postcombustion capture is the preferred approach for CCS in natural gas electricity production. Current postcombustion designs generally utilize a costly chemical separation process to "scrub" the flue-gas waste stream, but there is active ongoing research in membrane technologies as a potential alternative to improve the economics of this process.[40]

---

[38] NGCC plants employ a combustion turbine in series with a steam turbine. The hot exhaust from the combustion turbine raises the pressure and temperature of the steam for that portion of the plant, making it more efficient.

[39] National Energy Technology Laboratory, 2007a, 2010.

[40] In chemical separation, the gas mixture is dissolved in a liquid solution, usually an amine-based chemical in water, and the CO_2 is then removed from the mixture. Regarding membrane technologies, see Adhikari and Fernando, 2006, and Metz et al., 2005.

The institute could also develop alternative design concepts, if the aforementioned approaches appear to offer insufficient opportunities for cost reduction and design improvement. For example, the institute could explore precombustion capture to reduce or eliminate the technical issues associated with the CO_2-separation processes required by postcombustion approaches. One approach to precombustion capture is to initially convert natural gas into a synthesis gas (or syngas)—a fuel that consists primarily of hydrogen—and then combust the syngas.[41] In this case, CCS is applied to the precombustion waste stream, which has high concentrations of CO_2, rather than the flue gas, which is dilute and consists mostly of water as a waste product.

Another alternative design concept could be oxyfuel combustion (i.e., burning natural gas in a pure stream of oxygen rather than in air). In this case, the postcombustion CCS is able to take advantage of a concentrated waste stream, which again minimizes requirements for more-efficient and cost-effective CO_2 separation technologies. Both of these concepts have analogues in advanced coal-combustion technology and so are being researched elsewhere, but the concepts might be applicable to natural gas as well and should therefore be considered in that context at the institute. In considering research on alternative designs, the institute would need to first assess the theoretical costs and benefits relative to, for example, NGCC with CCS to determine whether these new research directions have potential merit.

One very important consideration in determining the best overall approach for CO_2 separation in Qatar will be water use. For example, use of the current commercially available chemical separation process would significantly increase water consumption above the already water-intensive operations at a wet-cooled thermal plant. Alternative chemical separation approaches that are less water intensive or dry cooling might therefore be necessary. Use of dry cooling can significantly increase costs, so this might not be an economically viable option.[42]

---

[41] Syngas consists primarily of hydrogen and carbon monoxide. The syngas can also be further processed in a water–gas shift reaction to produce nearly pure hydrogen and an additional stream of CO_2, if a separation process is employed.

[42] Zhai, Rubin, and Versteeg, 2011.

Precombustion capture or oxyfuel approaches might have some advantages in terms of water use, and this factor should be considered in weighing the relative benefits of these various approaches to enabling CCS.

There are numerous opportunities for synergies of CCS research with other recommended research areas. For example, natural gas with CCS could be further coupled with concentrating solar power to enable virtually zero-emission electricity production. If the institute develops GTL technologies that utilize a syngas approach, precombustion CCS would be applicable to reducing the GHG intensity of the process. Overall, there is incentive for the institute to advance CCS technologies that use natural gas, both for reducing Qatar's GHG emissions and to ensure a sustained high demand for natural gas around the world in a highly GHG-constrained future.

Collaboration Opportunities and Human-Capital Needs for Qatar

As noted, worldwide, there is limited research in natural gas with CCS because coal-based electricity dominates. Thus, there is a unique opportunity for Qatar to set the research agenda and lead in technological breakthroughs in this field. Already, Shell, Qatar Petroleum, and the Imperial College London are collaborating with QSTP to create a CCS research center. The institute should collaborate closely with this center and develop a complementary research agenda in CCS.

In the GCC, there are opportunities to collaborate with Kuwait University and King Saud University in KSA, which are undertaking research in CO_2 conversion to materials with economic and environmental benefits. UAEU in Al Ain is developing methods of removing CO_2 from natural gas using membrane contactors and of using CO_2 in oil recovery. UAEU in particular has expressed interest in collaborating with Qatar in its CCS research.

There might be opportunities to collaborate abroad, as well as opportunities to export natural gas with CCS technology. For example, in the United States, the state of California is interested in applying CCS to natural gas due to its own relatively large use of natural gas–fired electricity production. This could present a technology-export opportunity, a collaboration opportunity, or both.

To assess the CCS potential in Qatar and the GCC region, the institute would require geologists and engineers (e.g., mechanical, civil). Development of natural gas–specific CCS would require electrical and mechanical engineers with specific expertise in electrical power systems and natural gas turbines. There is some potential to develop these skills within the GCC. King Fahd University of Petroleum and Minerals in KSA, for example, offers civil engineering Ph.D.'s with a focus in geotechnical engineering. In addition, the UAE offers undergraduate courses of study in applied geology and petroleum geology leading to M.Sc.'s in petroleum science and engineering or remote sensing and geographic information systems.

Priority Topic 4: Developing Solar Energy

Background and Motivation

Solar energy is one of the only major renewable resources in Qatar, but it is plentiful there and in the region. As such, it is a potentially major resource for Qatar's long-term energy security. If even a small fraction of the energy in incident solar radiation could be economically collected and put to productive use, all of Qatar's energy needs could be met by sunlight alone.[43] As noted by the chief executive officer (CEO) of Abengoa Solar, the Spain-based global solar company, "Northern Africa and the Middle East are undoubtedly areas with a tremendous solar energy potential, for both the region's own use as well as exporting."[44] Qatar should aim to position itself as a research leader in this solar future.

Solar energy can be used in several ways. It can be captured as heat (called solar thermal energy) and used to heat or cool water or

---

[43] If all the sunlight that reaches the earth's surface could be captured and productively used, the harvested energy would exceed human energy needs by nearly four orders of magnitude (U.S. Department of Energy, 2005).

[44] Appleyard, 2010a. Export of solar electricity from the GCC region will require significant infrastructure investments, including a subsea power line from, for example, the MENA region to continental Europe. Some direct solar fuels, however, might be more readily exported with existing or incrementally modified infrastructure.

ambient air in buildings. It can be converted to electricity and directly used or supplied to the electricity grid for any number of applications. It can also be directly converted into chemical energy (i.e., a fuel). One important application of these processes is desalination, either via direct thermal energy or via indirect electrical or chemical energy.

These processes—heating and cooling, electricity production, combustion of fuels, and desalination—typically rely on fossil fuels, which emit significant GHGs and are nonrenewable. If solar energy could be economically substituted for or used to supplement fossil fuels, significant GHG emissions could be avoided. Qatar (and much of the rest of the world) would have an abundant source of clean, renewable energy well into the future.

However, many kinds of solar energy conversion are presently too expensive and are, as yet, technologically insufficient to realize this full potential.[45] Even in cases in which the technologies are already competitive with traditional fossil-based ones in some parts of the world (e.g., domestic solar water heating), there are often market or logistical barriers to deployment of the technology that could be reduced by applied technology and policy research.[46]

We recommend that the institute consider research in five key areas that address critical needs and opportunities in Qatar:

- concentrating solar power (CSP) hybridized with natural gas combustion
- space and water heating and cooling
- solar photovoltaics (PV)
- direct solar fuels
- solar desalination.

We recommend that the institute undertake CSP–natural gas hybridization, space and water heating and cooling, and solar-desalination research as core elements of its solar research program and that it consider PV and direct solar fuels as secondary research. We

---

[45] Crane et al., 2011.

[46] Western Governors' Association, 2006.

discuss four of these topics here but present solar desalination as part of a broader discussion of desalination research (topic 9).

Table 4.3 summarizes the commodities that each of these solar technologies provides. Collectively, they produce electricity, fuel, commodities that are presently made with electricity or fuel (e.g., building heating and cooling), and water. Electricity, transportation fuels, and water are all required for a modern standard of living. They are also independently relevant to sustainability and fundamentally linked.

Recommended Research: Concentrating Solar Power Hybridized with Natural Gas

Electricity in Qatar is almost entirely produced with natural gas. Although this is an abundant regional resource and a relatively clean fossil fuel, it is ultimately a finite resource with significant GHG emissions. CSP technologies use solar energy to produce electricity and, when hybridized with natural gas combustion, offer many advantages. CSP hybridized with natural gas would extend the life of Qatar's natural gas resource, reduce the emission intensity of electricity production, and provide flexibility in electricity production that is not possible with CSP technologies alone. As such, this technology is perhaps the perfect match for Qatar's two most-abundant natural resources. This should, accordingly, be a high-priority research area for the institute.

Table 4.3
Solar Technologies, Initial Conversion Mechanisms, and Delivered Products

Technology	Initial Solar Conversion	Delivered Commodity
CSP hybridized with natural gas	Thermal	Electricity
Residential, commercial, and industrial space and water conditioning	Thermal	Heating and cooling
PV	Electrical	Electricity
Direct solar fuels	Chemical	Fuel
Solar desalination	Electrical or thermal	Water

Thermal power plants use steam to drive a turbine and produce electricity. In Qatar, natural gas is combusted to produce this steam, but it could instead be produced entirely or in part by solar thermal energy. As the name suggests, concentrating solar power uses optical lenses and mirrors to concentrate sunlight to obtain high temperatures, usually by heating a working fluid to high temperatures.[47] The heat is then transferred to produce high-pressure, high-temperature steam (sometimes over 700 degrees C). CSP requires high-intensity, direct solar radiation, so it is generally most appropriate in hot, relatively cloudless regions; its range of practical implementation is therefore more geographically limited than some other solar technologies, but it is likely to be appropriate throughout the GCC countries.[48]

Yet, even in ideal environments, CSP is not well suited to be the only source of steam in a baseload power plant because solar energy is intermittent.[49] However, CSP and fossil fuels can be hybridized to provide baseload power while also reducing fuel use and GHG intensity relative to use of fossil fuels alone. Hybridization can be accomplished in at least two ways. Solar energy can be used as the primary source of electricity, with the other steam-producing fuel as a supplement or backup. Alternatively or additionally, solar thermal energy can be used to supplement another primary source of heat, such as natural gas combustion, either before or after the steam is generated by the primary

---

[47] This working fluid needs to be resistant to breaking down at high temperatures. For example, in some designs, the working fluid is silicone oil.

[48] A detailed assessment of the solar resource in Qatar and the GCC region would enable this to be verified and quantified. This would be part of an energy-resource characterization, which we recommend in topic 8, energy strategy.

[49] Note that, in Qatar, peak demand might be driven by electrical air-conditioning needs, which are highest in the middle of the day. This coincides with the availability of solar energy, suggesting that solar energy could be well suited to address peak demand. CSP can also be coupled to thermal storage, e.g., large reservoirs of molten salts, to enable more-consistent electricity supply, including electricity production at night. In this case, the molten salt is heated when the sun shines during the day and then is used to heat steam when the sun is less intense and at night. However, due to Qatar's abundant natural gas resource, and due to interest in thermal storage and ongoing research in this area in other parts of the world, we do not recommend this research area for the institute at this time.

fuel, depending on the system design. This reduces the consumption of primary fossil fuel.[50]

The institute should research CSP technology and, in particular, optimizing the design of truly hybrid CSP/natural gas power plants, in which solar energy is used to supplement natural gas. CSP designs differ primarily in their optical layout. Two CSP technologies dominate in other parts of the world, and these two, among others, should be considered for use in Qatar. First, parabolic-trough CSP designs are most mature and are the dominant technology in the United States. These designs use parabolic mirrors to concentrate solar energy on a tube of working fluid.[51] Second, central-receiver technologies are increasingly being commercially deployed, especially in sunny regions of the European Union (e.g., Spain).[52] These designs concentrate energy on a single point. The institute should undertake research to select an appropriate CSP design for Qatar. It can also research improvements to optics and optical configurations (i.e., the mirrors and their particular orientation for focusing and concentrating sunlight) in a specific design. The choice of design, configuration, and location for Qatar could be based on a combination of its climate and on technical synergies with existing electricity supply technologies in Qatar.[53] There are multiple CSP

---

[50] Natural gas backup of CSP plants has been around for decades (see, for example, the U.S. solar-electric generating station [SEGS] plants listed at National Renewable Energy Laboratory, 2010b). Truly hybridized systems, however, are not widely commercialized.

[51] The first commercial "hybrid" CSP/natural gas facility has recently come online, making use of parabolic trough technology (see Olson, 2011). However, this system was originally an integrated gasification combined-cycle (IGCC) plant, and the CSP component has been added on in a retrofit. The system was not designed from the ground up as a CSP/natural gas plant. The research we are recommending herein would be truly innovative, from-the-ground-up CSP/natural gas hybridized systems, optimized for application in the GCC region.

[52] There are multiple basic CSP designs with a range of technical maturities, including parabolic trough, central receiver (also known as a power tower), linear fresnel, and parabolic dish (e.g., Stirling engines, which eliminate the heat-transfer step by directly using the working fluid) (National Research Council, 2010).

[53] For example, humidity reduces the efficiency of CSP. Humidity can be high in Qatar near the coast but can decline as one moves to the interior. This suggests that CSP could be more efficient away from the coasts but that siting away from the coast could increase electricity

designs and technical approaches, so there are likely to be opportunities for both near-term and high-risk research.[54]

There might be additional opportunities to optimize or modify existing technologies to tailor them specifically to use in Qatar and the GCC region. Near-term, low-risk research might include developing more-robust heat-transfer fluids, potentially with a dual use for thermal storage. The institute could also research more–thermally robust and efficient or less-expensive optics and optical coatings. More high-risk, disruptive research might involve complete redesign of a CSP system rather than making incremental improvements to more-mature concepts, or it could involve development of new ways to couple CSP and natural gas combustion.[55]

CSP has strong synergies with some other priority research areas that we recommend. For example, combining CSP, natural gas, and CCS technologies could enable near-zero GHG emissions from electricity production and extend the use of the finite supply of natural gas. The use of CSP also has implications for water use and desalination. For example, research in improving the efficiency of dry-cooling systems, which use much less water than traditional cooling systems, would enable the more-widespread deployment of CSP in particular and increase the sustainability of thermal electricity production

---

and water transmission and distribution costs. System research would be important in identifying siting that makes the best trade-off between these concerns, given Qatar's particular climate and geography.

[54] One specific design for a high-efficiency CSP/natural gas hybrid is an integrated solar combined-cycle system (ISCCS). This concept integrates a parabolic-trough plant with a conventional NGCC plant. NGCC plants employ a combustion turbine in series with a steam turbine. The hot exhaust from the combustion turbine raises the pressure and temperature of the steam for that portion of the plant, making it more efficient. When hybridized with parabolic trough, solar heat is used to further increase the temperature of the high-temperature exhaust from the combustion turbine and improve the performance of the steam turbine (National Research Council, 2010). The first two ISCCS plants are being built in Morocco and Algeria.

[55] One potential concept is direct steam generation in which the solar energy is used to directly boil and heat water rather than using an intermediate heat-transfer fluid. This design would offer potential increases in efficiency. Regarding new ways to couple, see National Renewable Energy Laboratory, 2010a, and National Research Council, 2010.

in general. This is because, if dry cooling were used, thermal plants could be sited away from water sources and would minimize the use of water resources entirely. CSP research could also complement solar-desalination research by providing lower-grade waste heat from electricity production to perform or enhance thermal desalination.

Recommended Research: Solar Thermal Space and Water Cooling and Heating

Buildings consume a substantial amount of energy, and heating and cooling, in turn, make up a significant portion of that total energy use.[56] Solar thermal energy can be used to heat or cool ambient air or water, replacing fossil fuels that are traditionally used for such applications. For example, to reduce overall electricity demand in buildings, solar air conditioning can leverage the solar resource when it is most needed, at the hottest part of the day, when air conditioning use is high. Solar thermal technology and policy research should be another high-priority subtopic for the institute: Substantial new building construction is occurring in Qatar, and it presents a good opportunity for incorporating new building technologies.

Space and water conditioning are thermal technologies but make use of much lower temperatures than, for example, CSP (~50–400 degrees versus more than 700 degrees C). These technologies include passive heating and cooling of ambient air; direct solar water heating for domestic and industrial uses, with storage in insulated reservoirs; and thermally driven solar sorption (i.e., absorption or adsorption) chillers for air conditioning.[57] Similar sorption concepts can also

---

[56] For example, in the United States, around 40 percent of total energy is used by commercial and residential buildings, and heating and cooling account for approximately 40 and 70 percent of this total energy, respectively (Energy Information Administration, 2010).

[57] Passive heating and cooling make direct use of the sun's thermal energy to collect, store, distribute, and reject heat in buildings. They do not make use of any mechanical or electrical devices in their use of sunlight; those are considered the active components of other solar utilization technologies.

Some solar chillers use a low-boiling-point refrigerant, which consumes heat to evaporate, so the cooling effect can be used to condition spaces in buildings. The refrigerant is condensed back into a liquid in a thermal process using the solar heat. This is in contrast with the more-

be used in solar refrigerators for homes and solar process coolers for industrial applications, other ways in which solar energy could replace the traditional use of electricity.[58]

In some cases, solar space- and water-conditioning technologies are relatively mature or cost-effective compared with some other solar technologies, in part because they harness approximately 60 to 80 percent of the incident solar energy, and they can be used for domestic or commercial applications. As such, they are likely to be able to offer significant reductions in the GHG intensity of heating and cooling in Qatar in the near term. Additionally, depending on the specific technologies, implementation is often relatively simple, and operation and maintenance (O&M) is often minimal.

The institute should undertake technology research that further improves the efficiency or reduces the costs of these technologies or optimizes them for use in Qatar's environment. For example, research in improved refrigerants and desiccants or thermally robust, inexpensive materials for these lower-temperature applications might be appropriate. In some cases, relatively incremental efficiency increases and cost decreases, rather than disruptive technology changes, might be expected and would likely be sufficient for widespread deployment on the basis of technology readiness.

The institute should also undertake policy research to determine the range of applications for solar space and water conditioning in Qatar, estimate the potential energy and GHG savings, and establish how best to accelerate their adoption through financial incentives, regulations, or other policy levers. This research would be highly synergistic with green building research—solar thermal systems can be integrated with other green design elements, and more-efficient residential and commercial buildings would decrease the demands for air and water conditioning.

---

common electrically driven air conditioners, which use mechanical compression to condense a refrigerant. Other cooling systems use desiccants to facilitate evaporative cooling, and the desiccant is regenerated with solar drying.

[58] Appleyard, 2010b.

Recommended Research: Solar Photovoltaics

Solar PV technologies convert sunlight directly into electrical current via the photoelectric effect. Although the costs of PVs have declined steadily in the past few decades, electricity generated by PV still remains costly relative to almost all other forms of electricity production.[59] PVs also suffer from intermittency (i.e., they do not produce electricity consistently when sunlight is obscured by cloud cover, or at all at night). Therefore, PV is currently not economically competitive for bulk power generation, although it can be in niche and off-grid applications, such as solar-powered traffic signs or in remote locations. PV therefore presents a smaller opportunity to produce energy and reduce GHGs in Qatar, at least in the near term, than the various types of solar thermal energy discussed earlier. Other institutions and countries are also heavily researching PVs. In sum, PV research is likely to be less critical for Qatar than other solar energy research areas.

Nevertheless, the institute might wish to conduct some PV research to supplement a core solar energy research program. Such research could provide important opportunities for collaboration with the international solar research community. For virtually all types of PV, there are two major research challenges that need to be addressed. First, there is a need to increase the efficiency with which materials can capture solar energy and convert it into electricity without making the PV still more expensive. Second, PV production itself is very costly, and research is needed to develop manufacturing processes that reduce this cost without sacrificing the efficiency that has already been achieved.[60]

Additionally, Qatar should consider research to optimize PV in the Qatari climate—for example, with coatings that are optically transparent but still resistant to sand and pollution, or coatings that can handle extreme temperatures. One approach might be to assess the broad range of PV technologies under development and identify only one or a few with particular promise for Qatar. For example, concentrating PV (CPV) requires solar conditions much like those needed for CSP and therefore is especially suited to desert climates and would

---

[59] National Research Council, 2010.

[60] National Research Council, 2010.

offer important overlap with CSP research. On the other hand, the ability of thin-film PV to perform well with diffuse light would not be an advantage of great importance in Qatar. Building-integrated PV designs could have important applications in and synergies with green building design and deployment.

Recommended Research: Direct Solar Fuels

Solar energy can also be harnessed as chemical energy in the form of a fuel, as is done by photosynthetic bacteria and plants in biological systems. Indeed, many schemes for directly producing fuels from sunlight involve modified natural organisms or are inspired by natural design concepts and are performing what is accordingly often referred to as *artificial photosynthesis*. Given the abundance of fossil fuels, such as natural gas and petroleum, in Qatar, direct production of solar fuels is not a near-term need in Qatar. However, natural gas and petroleum are ultimately finite resources, and solar fuels could replace them in the longer-term future. Additionally, direct solar fuels would allow Qatar to export a natural resource—in this case, sunlight—in more-traditional ways than thermal or electrical solar energy could be stored and transported. As such, this research area should be considered as a potentially promising long-term, high-risk direction for the institute.

Direct solar fuels can include a range of technical approaches to yield different potential fuels.[61] Often, photosynthetic microorganisms are genetically modified to produce a fuel of interest, such as a liquid hydrocarbon that could directly replace petroleum or another fuel entirely, such as hydrogen. Ideally, the fuel would be nondestructively harvested, and the organisms would continue to efficiently produce fuel, making use of abundant nutrients.[62] Another approach is the use of synthetic chemical systems that mimic photosynthesis in a reaction vessel to produce the solar fuel. Note that producing direct solar fuels

---

[61] U.S. Department of Energy, undated (a), undated (b).

[62] Other approaches require that the fuel be extracted in such a way that the organisms would not survive the process, and new organisms would need to be cultivated to produce more fuel. This is generally the case with many precommercial technologies involving algae for fuel.

as presently conceived is generally water intensive, so this would be a potential risk in undertaking this research unless alternative concepts could be developed.

The institute's research in direct solar fuels would be focused largely at the exploratory, basic end of the research spectrum. It will be important to significantly increase conversion efficiency relative to natural photosynthesis—which is a small percentage or less in most organisms—for the concept to ever become commercially viable. Additionally, as noted, water use is a major concern for these technologies in general and in Qatar in particular, so research at the institute would need to focus on organisms that tolerate salt or are able grow in non-potable water, or it would need to develop an entirely new low-water concept.

Crosscutting Research

Some broad research areas might benefit many solar energy technologies. For example, more fundamental research in heat transfer could improve CSP, solar water heating, and CPV, as well as thermal solar desalination. The development of heat-, ultraviolet (UV)–, and dust-resistant plastics could be applied to virtually all solar technologies. Sandstorms and other environmental conditions can affect the output of solar energy technologies, so research on siting that takes into account sandstorms, humidity, and other factors would be valuable.[63] There could therefore be value in pursuing research that spans multiple solar subtopics, and a broad portfolio of solar-technology research that includes all or most of these topics could leverage synergies that otherwise would not exist.[64] This includes breadth in energy-portfolio analysis and policy research, as well as more-traditional bench science and engineering.

---

[63] As solar energy use grows, research to predict sandstorms and anticipate changes in energy supply and demand would also have merits.

[64] It might also be appropriate to engage in more-basic or general research and development, and synergies might be expected from a robust research program that spans basic to applied research.

Collaboration Opportunities and Human-Capital Needs for Qatar

Most GCC countries are undertaking a wide range of solar energy research, so there are many regional collaboration opportunities for the institute, from basic research to technology development and deployment to policy analysis.

It is important to note that, because Qatar has extensive reserves of natural gas, as well as solar energy, it is well positioned to lead in CSP and natural gas hybridization research in particular, though other GCC countries are also working in this area. For instance, the largest CSP plant in the world is expected to be built outside Abu Dhabi and to use natural gas to supplement solar energy.[65]

Most research institutes in Qatar are also conducting some form of solar energy research. GreenGulf (in collaboration with Chevron) has a site reserved for a comparative study on the efficiency of different solar technologies in Qatar. Texas A&M, Qatar University, College of the North Atlantic–Qatar, and Qatar Foundation are all currently conducting solar energy research, focusing on PV and smart grid integration.

Regionally, King Saud University in KSA and KISR in Kuwait are focusing on smart grid integration of solar energy, while the Masdar Institute of Science and Technology, KFUPM, and KAUST are researching solar applications in buildings. In the greater MENA region, the Egyptian Solar Research Center and the University of Jordan are conducting work in CSP applications.

Internally, there will be also important synergies with other research at the institute. For example, research in advanced low-energy buildings and smart grids would need to take the institute's emphasis on solar, and the intermittent nature of solar-based electricity, into consideration. The nature and magnitude of power fluctuations will be very different if making use of CSP/natural gas hybridization than if large-scale solar PV were deployed. Rooftop PV deployment would have implications for local grid fluctuations and would imply research needs for smart grids and small-scale backup power in buildings.

---

[65] Chadha, 2010.

If all of these topics in solar research are pursued in full, a relatively broad range of expertise will be required. For example, to support efforts in CSP, the institute would require expertise in several engineering disciplines (e.g., electrical engineering, chemical engineering, materials science), as well as physicists and chemists. To conduct research in direct solar fuels, plant biologists and microbiologists would be required. With the ongoing research at universities and institutions within Qatar, there is already a broad range of expertise in these fields. The prospect of pilot-testing sites, such as the Chevron/GreenGulf collaboration, along with a dedicated research facility could draw in more talent from around the globe. Regionally, Qatar could look to recruit from such institutions as the Masdar Institute of Science and Technology, KFUPM, KAUST, and UAEU, which all offer advanced degrees in some of these fields, as well as the potential to develop a research focus in solar applications.

Priority Topic 5: Fuel Cells

Background and Motivation

Fuel cells are electrochemical devices that combine a fuel and oxygen to produce electricity. Although some fuel cells can run only on hydrogen, others can run on alcohol or fossil fuels, including natural gas. It is important to note that fuel cells produce electricity more efficiently and cleanly than combustion of fuel.[66] Therefore, fuel cells that use natural gas are promising for Qatar's long-term sustainability.

High-temperature, stationary fuel cells can be used to supply electricity to an individual building or facility or to the electricity grid. Such fuel cells could offer some important benefits over traditional, combustion-based electricity production. First, they can generate electricity much more efficiently than combustion-based methods because they convert energy in a direct chemical process rather than in an indi-

---

[66] Minh, 2004; Jacobson, 2010.

rect thermal process.[67] Second, more-efficient fuel use means that fewer GHGs are emitted per kilowatt-hour of electricity generated. Third, stationary fuel cells use much less water than conventional electricity plants do.[68] Fourth, fuel cells have relatively concentrated CO_2 waste streams, which would make CCS easier to implement. Finally, if fuel cells are used to generate electricity at or near the use site, electricity transmission losses can be reduced or virtually eliminated, and even greater efficiency overall can be achieved. This distributed energy production could also increase energy security at the site.[69]

Thus, Qatar's natural gas resources would be used much more efficiently at home to produce electricity with fuel cells. Additionally, the combination of natural gas and fuel cell use, much like combining efficient natural gas combustion with CCS technologies, can make Qatar's natural gas resource still more valuable in a CO_2-constrained world.

Although fuel cells have much promise, they are not yet economically competitive with traditional combustion for electricity generation. Existing fuel cells that can run on natural gas are also small—frequently less than 10 megawatts (MW). In order to replace or augment traditional combustion, they must also be scaled up to hundreds of megawatts. Additionally, the efficiency of today's fuel cell technologies is comparable to today's high-efficiency combustion plants. Thus, it is likely that significant gains will need to be realized before the potential efficiency of fuel cells can be realized and before fuel cells can provide bulk power at an acceptable price. This suggests that fuel cells are a

---

[67] A fuel cell efficiently converts the chemical energy in fuel directly into electricity, rather than indirectly via the intermediate step of heating a gas to obtain mechanical motion in an electric turbine or automotive piston. The theoretical efficiency of fuel cells can approach 80 percent or even higher when making use of the waste heat. In practice, although fossil fuel–driven electricity plants run at efficiencies around 30 percent, high-temperature fuel cells can approach 60–70 percent efficiency (Stambouli and Traversa, 2002; Winkler and Lorenz, 2002).

[68] Traditional thermal plants generally require high-temperature, high-pressure steam to operate, and they use substantial amounts of water for cooling. Fuel cells use much less water. Moreover, this water can be recycled more easily and need not be lost via evaporative cooling.

[69] National Energy Technology Laboratory, 2007b; Grol, 2009.

way for Qatar to improve electricity production in the longer term, rather than in the near term. We recommend that the institute focus on technology research to reduce fuel cell costs and to increase power to many hundreds of megawatts without sacrificing the reliability and life span of fuel cells.

Recommended Research: Improving Solid-Oxide Fuel Cell Cost and Performance

The fundamental reaction in a hydrogen fuel cell combines hydrogen and oxygen to generate water.[70] In devices that can run on other fuels, such as natural gas, the hydrogen comes from the fuel, and CO_2 is produced as a waste product in addition to water. Multiple fuel cells are combined into fuel cell stacks to generate the desired level of power.[71] The stack is combined in a complete system, and the term *balance of plant* refers to all the other components of the system, including heat-management materials and structure and interconnections (e.g., wiring).

Fuel cells are generally characterized by the electrolyte they use for isolating the two halves of the cell in which the separate chemical reactions occur. Solid-oxide fuel cells (SOFCs) (which therefore use a solid oxide as the electrolyte) are used for natural gas–based fuel cells for two reasons.[72] First, they can use a broad range of fuels, including natural gas. Second, they can (and must) operate at high temperature,

---

[70] All fuel cells have the same basic structure and fundamentally consist of (1) an anode, where the catalyst active material breaks down the fuel to form hydrogen ions, free electrons, and waste in the case of nonhydrogen fuel (e.g., CO_2); (2) the cathode, where oxide ions are created and combined with hydrogen ions to form water; (3) the electrolyte, which separates the anode and cathode and selectively allows only the appropriate elements (as ions) to pass and does not allow electrons to pass. The free electrons generated at the anode are forced to move through an external load to complete the circuit before they recombine with ions at the cathode, and this process produces electricity.

[71] Combination of cells in series increases voltage; parallel cells increase current. Increasing the surface area of an individual cell also increases current.

[72] In SOFCs, nickel or other metals serve as the catalyst, and the solid-oxide electrolyte is most commonly an yttria-stabilized zirconia (YSZ) material, a very durable solid. The electrolyte is conductive to ions only at high temperatures (around 950 degrees C). This material presents advantages and disadvantages.

which has benefits for use of nonhydrogen fuels.[73] At present, SOFCs are generally viewed as the high-temperature fuel cells of choice in the near to medium term for stationary applications.[74]

However, there are many challenges to using SOFCs for large-scale electricity production, which we recommend that the institute address. First, the high temperature of SOFC operation presents challenges: Only high-cost interconnects (which link the various parts of the fuel cell stacks together) can withstand those temperatures.[75] Electrolytes that can operate at moderate temperatures are actively being sought, as well as more-robust interconnect materials. Second, most successful SOFC designs are for relatively small (<10s of megawatts), distributed units that generate power (and heat) on-site where needed. They have been deployed in many demonstrations and in cases in which there is a premium placed on reliability (e.g., off-grid or critical backup applications).[76] Third, although SOFCs are commercially available, they are expensive. Electricity from SOFCs costs about $4,000–

---

[73] High temperatures allow the electrolyte to work efficiently. High temperatures also make it possible to minimize "poisoning" of the catalyst (the actual material that breaks apart the fuel to form hydrogen ions), which can occur in the presence of contaminants associated with nonhydrogen fuels. Alternatives to YSZ are presently being researched that allow lower operating temperatures but maintain functionality and resistance to poisoning.

[74] These high-temperature fuel cells have been developed primarily for use in stationary applications. Note that we do not recommend research in polymer electrolyte membrane (PEM, also known as *proton exchange membrane*) fuel cells at this time because these are more-likely candidates for mobile applications. These must operate at relatively low temperatures to avoid damaging the plastic-based electrolyte. Generally, these are envisioned as hydrogen-powered devices, for two reasons: (1) conversion to a "hydrogen economy" for vehicles would require substantial investments in new infrastructure and substantial technical breakthroughs in storage technologies, and (2) Qatar has sufficient amounts of petroleum, or, alternatively, natural gas, to power its vehicles, so this type of transition is not likely to be a near-term priority in the region. This would be a potential future direction for the institute should circumstances change.

[75] Note too that, at high temperatures, some unwanted side reactions can occur, which can cause buildup in the cell and prevent proper movement of materials.

[76] SOFCs can also be used for mobile applications, but, most often, these are being considered for large transport vehicles rather than passenger vehicles due to the issues with heat management.

5,000 per kilowatt. Costs must be reduced to $400–500 per kilowatt to be truly competitive with bulk power.[77]

In order to realize the promise of fuel cells for Qatar's natural gas resources, the institute should conduct research to lower costs per kilowatt and to increase power to many megawatts, without sacrificing reliability and life span (goals that are typically at odds). This includes improving the performance of a single fuel cell (e.g., by developing materials that allow higher voltages across a single cell or by increasing the practical current density of materials).

It also includes scaling up the system to ever-larger stacks of linked fuel cells and larger blocks of stacks (e.g., through system engineering and better thermal management), without further increasing the costs of fuel cells. SOFCs are modular and are therefore very scalable even without increasing stack size—from several kilowatts to hundreds of megawatts. Thermal management and system integration, however, becomes a more important issue when scaling up plant size.[78]

The institute should also conduct research aimed at cost reductions by, for example, the use of lower-cost materials. Costs could be reduced either with robust but low-cost interconnects or by reducing temperatures without losing efficiency so that low-cost interconnects could be used. In particular, electrolyte alternatives that enable good performance at lower temperatures would be desirable. Generally improving the robustness of materials to allow long cell lifetimes with minimal degradation would also be a worthy research goal.[79] Further-

---

[77] U.S. Department of Energy, 2011.

[78] Organic Rankine cycle engines use waste heat to produce electricity. They use a low-boiling-point organic solvent, rather than water and steam, and are therefore applicable to lower-temperature thermal resources, such as geothermal or solar thermal turbines. However, they can also be employed to make use of the lower-temperature waste heat from higher-temperature thermal technologies, including natural gas combustion and stationary fuel cells. In this case, this can further increase the efficiency with which fuel cells produce electricity and be part of a thermal management approach.

[79] An additional area of research worth considering is developing in situ reformation of fuels—i.e., cleaning up or otherwise processing fuels at the site of the fuel cell, rather than transporting ready-to-use fuels.

more, although water use is low relative to combustion-based electricity generation, water usage could be driven even lower.[80]

There are many synergies between fuel cell research and other priority research topics we recommend. For example, hybridization of fuel cells and combustion turbines that run on natural gas are also possible and are a potential area of research for Qatar.[81] This would create strong synergies with natural gas research. There are also potential opportunities for overlap with concentrating solar power research and with CCS concepts that integrate fuel cells to reach virtually zero emission levels.

Collaboration Opportunities and Human-Capital Needs for Qatar

Fuel cell research and natural gas research are complementary and offer collaboration opportunities within the institute. The institute can collaborate with others within Qatar, such as the Fuel Cell Laboratory at Texas A&M University at Qatar, which is researching fuel cell–powered air-conditioning systems and grid integration.

Institutions in other GCC countries are also researching fuel cells. This includes KFUPM in KSA, which, in part, focused on developing fuels for fuel cells and mobile fuel cells,[82] and the Masdar Institute of Science and Technology, which is seeking to advance the state of the art in fuel cell design for a wide range of applications.

The institute can also collaborate with institutions abroad, including multinational corporations (such as General Electric, United Technologies, and Siemens) that are seeking to commercialize fuel cells, and research organizations (such as the U.S. Department of Energy's National Energy Technology Lab or the Pacific Northwest National Laboratory).

The institute would need to hire chemical and electrical engineers, materials scientists, and chemists and physicists, depending on the area of emphasis for the research. For example, improving thermal

---

[80] National Energy Technology Laboratory, 2007b; Tietz, 2007; Orera and Slater, 2010; U.S. Department of Energy, 2011.

[81] Winkler and Lorenz, 2002.

[82] Zaidi, Rahman, and Zaidi, 2007.

management is an engineering problem, but chemists and materials scientists are needed to develop new electrolyte materials. Both Qatar University (QU) and Texas A&M offer degrees and research opportunities in chemical and electrical engineering; however, there are no advanced degrees in these areas being offered in Qatar. Within the region, there are opportunities for advanced degrees in these fields. For instance, UAEU recently began offering a Ph.D. program in physics and engineering disciplines, and KFUPM has some established postgraduate programs in these areas as well.

Priority Topic 6: Green Buildings

Background and Motivation

Qatar has seen the rapid construction of new buildings in recent years: According to the Qatar Statistics Authority, there was a 50-percent increase in the number of buildings in Qatar between 2004 and 2010.[83] The building and construction sectors present both challenges and opportunities for Qatar in promoting sustainable development.

The building and construction sectors consume large amounts of energy, water, and other natural resources and produce waste, GHG emissions, and air pollution. Globally, about 30 to 40 percent of primary energy used in developed countries is used for heating, cooling, and other purposes within buildings,[84] and one-third of energy-related GHG emissions are associated with the building sector.[85] In Qatar, nearly all of this electricity and energy use in buildings is provided by natural gas. Hence, energy use in buildings contributes significantly to Qatar's GHG emissions. Additionally, the integration of buildings with transportation and economic infrastructure can dictate social and environmental outcomes,[86] so the design and siting of buildings during Qatar's rapid growth can considerably affect the country's development.

---

[83] Mansour, 2010.

[84] Huovila et al., 2007.

[85] Ürge-Vorsatz et al., 2007.

[86] Huovila et al., 2007.

Green buildings (sometimes called *high-performance buildings*) seek to use energy, water, and natural resources efficiently throughout the building life cycle and will be key to Qatar's sustainable development.[87] This includes increasing the use of renewable energy, promoting energy- and water-efficient technologies and behaviors, using environmentally preferable building materials, reducing waste and toxins, improving indoor air quality, and promoting smart growth.[88] Green buildings have been associated with a healthier, more productive workforce[89] and often are in high demand among knowledge-industry firms.[90]

The use of green building techniques can help Qatar reduce its energy, water, and resource consumption, which is currently among the highest in the world.[91] By reducing domestic energy consumption, green buildings can also give Qatar more flexibility to use its natural gas resources for exports or in ways that yield the greatest social and economic benefit. Green buildings can also have lower life-cycle costs than those of traditional buildings.[92]

Green buildings will be increasingly important to Qatar in the coming decades as Qatar's population and demand for buildings grow. Additionally, climate change data suggest that temperatures could rise in the region, increasing the need for water and cooling.

---

[87] Ürge-Vorsatz et al., 2007.

[88] EPA, 2010e.

[89] Singh et al., 2010.

[90] Eichholtz, Kok, and Quigley, 2010.

[91] Qatar's per capita energy and electricity use ranked in the top five countries globally in 2005 (International Energy Agency, 2007). Qatar's high energy use is in large part because Qatar's industries are energy-intensive and because Qatar has a small population, but Qatar is also in the top third of countries in terms of per capita residential energy consumption (World Resources Institute, undated).

[92] Although some green buildings require additional 1–3 percent initial capital costs, green buildings typically realize lower operating costs over their life cycle than those of traditional buildings. For example, in California, increasing building costs by 2 percent to invest in green building technologies has translated to a life-cycle cost savings of more than ten times the initial capital investment. See, for example, Kats et al., 2003.

Growing concern about resources and global climate change have spurred a global growth in green building practices. Technologies and policies applied elsewhere can provide a foundation for developing green building practices in Qatar, but they must be tailored to suit the local climate, geographic, cultural, and economic conditions in Qatar.

We recommend that the institute address this need with three themes of research and focus on water and energy consumption in each. First, the institute should conduct research to improve green building designs and technologies for the region. The development of Qatar's green building code has begun this process. The institute could add to the body of research on minimal-cost efforts to improve passive cooling and other green building efforts in warm or arid regions in the Middle East and Asia.[93] In addition, the U.S. Department of Energy has a series of recommendations for energy-efficient building in various climates, including hot and dry environments; these can provide a good starting point for the geographic context in Qatar.[94]

Second, the institute should conduct policy research to determine the most-effective ways of encouraging the use of green buildings in Qatar. Third, the institute should integrate urban-planning and green building research to address sustainable communities and cities. Finally, the institute itself should serve as a model and a test bed for green buildings. We discuss these in turn.

Recommended Research: Improving Green Building Designs and Technologies

The institute should first evaluate and compare how different building elements contribute to energy and water consumption.[95] Because energy and water use are not widely metered, this work might need to begin largely with existing literature. The institute could also select a

---

[93] For examples, see Ali and Al Nsairat, 2009; Aboulnaga, 1998, 2006; and Vangtook and Chirarattananon, 2007.

[94] U.S. Department of Energy, 2008.

[95] For example, in hot climates, building cooling typically dominates energy use by residential and commercial buildings. (Ürge-Vorsatz et al., 2007). However, lighting, appliances, and water use can also consume significant energy.

small, representative sampling of buildings and install meters to collect detailed energy- and water-use data.[96] Surveys are another approach, but it is difficult to collect detailed, accurate information in this way.

The institute should then assess green building designs that address these issues and determine trade-offs between different options. For example, lighting-energy requirements can be reduced by using more-efficient fixtures, placing lights on timers, and orienting buildings so that more natural light is available. Different options vary in their effectiveness and in their applicability to new and existing buildings. Differences in cost, social acceptability, and the effect on other resources also make them differently feasible. An assessment of these factors would enable the institute to recommend best practices for green building design in Qatar.

The institute should also undertake complementary research in advancing the state of the art in green building designs and technologies and tailoring them to the region. Most recent efforts to reduce building energy use focus on relatively temperate regions in the United States and the European Union. The institute should research and develop modifications of these technologies and develop new technologies to suit high-temperature, arid regions. For example, roof and window overhangs were originally developed for temperate climates to minimize solar exposure in summer (to reduce cooling needs) while maximizing exposure in winter (to reduce heating needs). This climate, of course, does not apply to Qatar, but the institute could improve such passive solar energy designs to minimize heat uptake and maximize access to natural light all year round. Because of the trade-offs involved in these choices, the institute should take a systems approach in which the interactions between energy, water, and other resource use are all considered.

In the following subsections, we describe green design features that are most important for Qatar. Where applicable, we have identified specific research questions associated with those designs. Readers

---

[96] In the future, advanced utility metering and distribution monitoring (topic 7, on smart grids) could provide high-resolution data on energy use.

can continue to the next subtopic on encouraging green building use without loss of continuity.

Siting and Orientation. Building location and orientation play a significant role in energy use. In arid climates, reducing solar exposure and taking advantage of natural wind patterns can greatly reduce the energy required to achieve a temperate indoor climate.[97] Shading by nearby buildings or vegetation can also reduce solar heating.[98]

Although siting and location designs are not technology-intensive, the institute should conduct research to tailor them to cultural and geographic conditions in Qatar and the region. For example, siting decisions should consider local wind patterns, local norms about the distances between buildings, and cultural expectations as to how buildings should look and function. Also, building-siting decisions and design decisions are related and must be made together. For example, a building located at a site with high levels of solar exposure might not need a design that incorporates large windows for natural light.

In collaboration with local green building organizations, such as Barwa and Qatari Diar Research Institute (BQDRI), the institute can develop recommendations to help site new buildings to fit local streetscapes and provide desirable sun and wind exposure.[99] These recommendations can include guidance about design elements that are appropriate for different site characteristics.

Passive Cooling. The institute should pay particular attention to cooling technologies, such as passive cooling designs for new buildings. Passive cooling systems or low-energy ventilation systems help circulate cooler air with little or no use of conventional energy.

---

[97] Ürge-Vorsatz et al., 2007.

[98] Huovila et al., 2007.

[99] Barwa is a Qatar-based real estate company engaged in the acquisition, reclamation, development, and reselling of lands to establish agricultural, industrial, and commercial projects. It is also involved in the real estate administration and operations sectors. Barwa focuses on the domestic and international real estate development business, investments, hotel ownership and management, financial services, consulting, advertisement, and brokerage service, among others.

Simpler approaches include improving window placement and using high-albedo roofing coatings that reflect solar radiation and reduce cooling needs of buildings.

More-complex passive ventilation designs create pressure differentials to circulate air from high- to low-pressure areas, without the use of conventional energy.[100] It is also possible to draw in air that is cooler than the ambient outdoor air temperature by having the inflowing air run first through cool underground piping.[101] Utilizing high amounts of thermal mass and nighttime ventilation techniques can also reduce initial daytime temperatures in buildings and reduce the rate at which temperatures rise during the day.[102]

Because of Qatar's high temperatures for most of the year, it is unlikely that the passive cooling can replace active technologies, such as air conditioning. However, passive ventilation during the nighttime in tandem with energy-efficient air conditioners can significantly reduce energy demands for cooling.[103]

Active cooling can also be made more efficient. One approach is to reduce the active cooling required for incoming air by using out-

---

[100] This process can be driven through temperature differentials. For example, in a residential solar chimney system, air in the chimney is heated by the sun (in some cases, by painting the chimney black so it absorbs more solar energy). Hotter air rises, driving the air in the chimney up and out of the building. Simultaneously, air from inside the building rises into the chimney, and air from outside is drawn in to maintain a constant amount of indoor air. This allows for more-efficient ventilation without an active energy source (Aboulnaga, 1998; Ala-Juusela, 2004).

[101] Huovila et al., 2007; Ürge-Vorsatz et al., 2007.

[102] Thermal mass is physical bulk that requires relatively high heat inputs to experience a change in temperature. Therefore, incorporating thermal mass into buildings smooths temperature changes. For example, air in a building begins to warm in the morning. The thermal mass will warm more slowly because it requires a higher heat input per degree temperature change than the air requires. Thus, thermal mass can serve to absorb heat from the air and slow temperature increases. Examples of thermal mass include earth, concrete, or other materials with a high heat capacity that are used as building materials.

See also Ürge-Vorsatz et al., 2007.

[103] Da Graça et al., 2002.

flowing indoor air to cool the incoming fresh air.[104] Another approach is to use water, rather than air, as the cooling medium because it is more efficient at heat transfer.[105]

The institute can also research advanced active cooling technologies, such as geothermal heat pumps, desiccant-based cooling, solar thermal cooling systems, and high-efficiency traditional air-conditioning systems with variable-speed fans. As with building siting and orientation approaches, the institute should explore the trade-offs involved in each cooling approach within a Qatari context and determine whether the benefits are substantial enough to merit the costs.

Finally, because water is scarce in the region, the institute should assess whether the reduction in energy use by using water-based cooling technologies is larger than the energy required to produce the water. It might be possible to optimize the technology for the local resource constraints, perhaps by decreasing cooling power in return for decreasing water usage in the water-based cooling system. This is one example of the importance of a systems approach that considers life-cycle energy, water, and resource use.

Improving Building Envelopes in New and Existing Buildings. Effective building envelopes are key to reducing energy use in both new and existing buildings.[106] They prevent heat transfer between the internal and external environments. In hot climates, such as Qatar's, a well-designed building envelope prevents the transfer of outside heat to the inside of the building; in cold climates, it prevents inside heat from leaking out. In both cases, it reduces the energy required to maintain comfortable indoor temperatures.

---

[104] A device that allows for efficient transfer of heat between incoming and outgoing air is known as a *heat exchanger*.

[105] In such a system, buildings are cooled by circulating cold water through pipes, often situated in ceilings. Airflow, and its energy requirements, can then be reduced to the lower level required for ventilation purposes (Ürge-Vorsatz et al., 2007).

[106] A building's envelope consists of elements that separate the interior from the exterior, such as its roof, foundation, windows, and exterior walls and doors.

To achieve an effective building envelope, it is necessary to insulate walls and use high-performance windows and doors.[107] Weather-stripping windows and sealing other areas of likely air leakage has reduced energy consumption of existing buildings by 10 to 30 percent at relatively low cost.[108] Windows are especially important because they transmit heat at four to ten times the rate per unit area at which other elements of the building envelope transmit heat.[109] Double-paned windows with an insulating spacer reduce heat transfer from the outside. In addition, window glazes can reduce the intake of passive solar heat by 75 percent by reflecting long-ray solar heat energy while allowing light to pass through.[110] Research is ongoing to develop more-effective window glazes, and the institute could research regionally appropriate window glazing and design.

There are also several methods of reducing the absorption of solar radiation by roofs. Roofs can be painted with high-albedo coatings to reflect sunlight, or they can be covered with vegetation. The institute could consider researching climate-appropriate vegetation for such green roofs. Successful, climate-appropriate vegetation could also be used to minimize landscaping water use.

Energy-Efficient Lighting and Appliances. Energy-efficient lighting and appliances can also reduce energy use in new and existing buildings. Energy-efficient lighting, such as high-efficiency fluorescent tubes, compact fluorescent bulbs, and solid-state lighting, both reduces the direct energy load for lighting and decreases the heat generated by lighting that otherwise would add to the cooling requirements.[111] The institute could pursue research into advanced solid-state lighting.

Research can also address methods of encouraging more–energy-efficient building operations. Timers and sensors can be used to optimize lighting and cooling loads for different occupancy levels.

---

[107] Huovila et al., 2007.

[108] Bell and Lowe, 2000.

[109] Huovila et al., 2007.

[110] Ürge-Vorsatz et al., 2007.

[111] Ürge-Vorsatz et al., 2007; Creyts et al., 2007.

Dimmer sensors can be used to automatically set light levels based on occupancy and to reduce artificial lighting near windows when it is sunny.[112] Offices can use task-ambient lighting, in which hallway and other background lighting is low but additional light is available at workstations. Energy-efficient appliances can reduce operating-energy requirements but often have higher initial capital costs. Finally, proper maintenance of appliances is critical to energy-efficient operation, yet is often neglected.

The institute should also assess the impact of appliances and other end-use technologies that use natural gas as a substitute for electricity. Energy is lost in the electricity-production process, as well as in transmission and distribution. Thus, using natural gas as a direct source of energy could require less overall power input than generating electricity from natural gas and then using that electricity as the energy source. Such appliances as natural gas–fired air conditioners and refrigerators are established technologies, and these appliances often require lower maintenance levels than electricity-driven models to maintain peak energy-efficiency.[113] However, capital costs are typically higher than for electricity-driven appliances. The institute should research the current costs and benefits of these appliances, how these costs might evolve if demand for these appliances were to increase, and whether the benefits in energy-use reduction merit policies for encouraging their use in Qatar.

Distributed Energy Generation and Use. The institute should also evaluate technologies that generate energy at the building site and reduce dependence on nonrenewable energy.[114]

---

[112] Ürge-Vorsatz et al., 2007.

[113] Natural Gas Supply Association, undated.

[114] It is also possible to use fossil fuels for distributed energy generation—e.g., with micro-turbines that combust natural gas to produce electricity near the use site. This again reduces electricity transmission losses. If Qatar pursues this kind of distributed electricity generation, the green building research program could also develop ways of reusing the waste heat from combustion. This is typically done in cold climates, where waste heat is used to heat buildings through CHP. The institute could research ways of using distributed waste heat to cool buildings, possibly with processes similar to those used in solar refrigerators and solar chillers, discussed in topic 4 on solar energy.

For example, green building design can incorporate solar thermal systems to provide heated water for the building. Building roofs and walls can also provide a surface for solar PV. These can be traditional solar panels or building-integrated PVs (BIPVs), which also function as part of the building's exterior. The institute should research the progress of the various PV technologies and estimate other deployment costs, such as installation. It should also evaluate which types of buildings and sites would make the most-efficient use of this technology. If widespread PV deployment appears tenable, the institute should formulate its building-siting and design recommendations accordingly—for example, to include the shape and orientation of roof surfaces. Because distributed solar generation might provide more electricity than the building needs, the institute should also explore technology to best integrate distributed electricity generation with the grid. Topic 7, on smart grids, discusses this further.

Water Conservation. Green buildings in Qatar should also reduce water use. Water use can be significantly reduced with water-conserving fixtures.[115] New building projects can use gray-water systems, in which water that has been used for washing (dishwashing, laundry, bathing) can be recycled for toilet use or irrigation, thus reducing demand for freshwater.[116] In general, technologies to reduce water consumption in buildings are relatively well developed. In a society in which water use has not been highly restricted, as in Qatar, developing effective strategies to encourage the use of these technologies will be crucial. We address this further in the discussion of topic 11, on water demand management, and more generally in the section on the recommended policy research to encourage technology adoption.

Building and Construction Phase. Buildings consume resources when they are being constructed, used and operated, and ultimately demolished. The energy used during the construction phase is less than

---

[115] Key targets are showers and toilets, since such activities as bathing, flushing toilets, and washing dishes compose the majority of household water use in Qatar (Al-Mohannadi, Hunt, and Wood, 2003).

[116] There has been some study of gray-water use in the Middle East. See, for example, Al-Jayyousi, 2003, and Prathapar et al., 2005.

20 percent of a building's life-cycle energy use.[117] Thus, minimizing the use-phase energy requirements of buildings with better designs and operation, as described earlier, should be prioritized over reducing the energy and resources needed to construct the building.

Nevertheless, there could be some low-cost methods of reducing resource use during construction as well. For example, local, recycled, and reused materials can have lower manufacturing and transportation-related environmental impacts and lower life-cycle impacts overall. Reused and recycled materials can also reduce the impact of demolition. The institute should research such options as alternative building materials to reduce the energy and environmental impact of construction and demolition.

Behavioral and Technology-Adoption Research. Finally, the institute should evaluate which new green building technologies will be accepted by building occupants in Qatar, and how. This should shape its technology recommendations. For many of the green building approaches discussed earlier, little behavioral modification is required of the building user. A well-insulated building that is oriented and designed to minimize solar-radiation absorption will have a lower cooling-energy requirement, regardless of occupant behavior.[118] However, reducing energy draws from lighting and appliances can require user participation and acceptance.

If the institute's research suggests that behavioral change will be difficult, the institute could focus its research on technologies that reduce resource use yet require little behavioral change. This research can also inform methods of encouraging green building use, discussed next.

---

[117] Huovila et al., 2007.

[118] Another example is district cooling, which centralizes air conditioning for a neighborhood by generating and distributing cold water that can act as a building coolant. District cooling can reduce the energy requirements of cooling through economies of scale. And it displaces the responsibility for cooling-equipment maintenance from private individuals to a utility with greater ability and incentive to ensure that maintenance is performed properly. District cooling has been implemented in some neighborhoods in Qatar by Qatar Cool. See International Energy Agency, 2004.

Recommended Research: Encouraging Green Building Use

Green building designs must be implemented in order to realize energy, water, and resource savings. Many policies can encourage the use of green building designs and technologies. Building codes and regulations are "stick" policies that require the use of green building concepts, while tax incentives and market signals are "carrots" that encourage use.[119]

Whether these and other approaches are appropriate and successful depends on many economic and social factors. For example, Qatari nationals are not accustomed to paying for the full cost of water and energy use, potentially making new pricing systems difficult to introduce. It is also possible that demand for these resources will be less elastic in response to price changes than it has been in other countries.

We recommend that the institute conduct policy research to assess the effectiveness, feasibility, and appropriateness of different policy options given Qatar's energy and environment goals and given Qatar's economy and culture. Thus, the institute should also consider modifications that might make policies more effective and more socially and politically acceptable. The institute should also conduct cost-effectiveness analyses of the best candidate policies.

In the following subsections, we describe policy options that might be possible in Qatar. Where applicable, we have identified specific research questions associated with those policy options. Readers can continue to the next section on collaboration and human-capital needs without loss of continuity.

Regulatory Policies. One policy option is to implement a mandatory green building code that is applicable to all new buildings. This approach could be very effective in developing a stock of new green buildings. However, it will have little effect on existing buildings unless retrofits are also required to ensure that all buildings meet some minimum green building standards.[120]

---

[119] There are several implementation efforts under way in such regions as the United States, the European Union, and Australia (Ries, Jenkins, and Wise, 2009).

[120] Different standards might be appropriate for new versus existing buildings.

Building codes can be prescriptive or performance-based. Each type has its own trade-offs. Prescriptive codes specify which designs or technologies must be used in buildings, e.g., low-flow showerheads or low-flush toilets. A performance-based code specifies energy, water, and other performance targets that the building must meet, but it allows flexibility in how those targets are met. Such a code could require, for example, that buildings limit per-occupant water use to a certain threshold, a requirement that can be met in many ways, including through use of low-flow showerheads or low-flush toilets. Prescriptive codes are typically easier to implement because it is easier to determine whether a building meets requirements than it is to measure performance. However, they can limit technological creativity. In contrast, performance-based approaches encourage technological creativity but can be difficult to implement because it is more difficult to determine whether the performance criteria have been met than it is to simply set requirements.[121]

In addition to building codes, regulators can ban energy inefficient technologies when affordable replacements exist. For example, the European Union, Australia, and others have begun phasing out the use of incandescent light bulbs.[122] In cases in which replacement technologies exist, the institute should conduct a cost-benefit analysis comparing energy-efficiency technology standards and bans on obsolete technology.

Market-Based Approaches. Market forces can incentivize green building use, in addition to or instead of regulations. For example, if building operators (or homeowners) pay for resources, such as water and electricity, then green building technologies that reduce the use of these resources can also save the consumer money in the long run. In Qatar, energy and water are available freely or at minimal cost, so such market forces are effectively nonexistent. The institute should research options for implementing an efficient and equitable market for electricity and water. Topic 11, on water demand management, further explores the role of pricing for reducing water consumption.

---

[121] Ries, Jenkins, and Wise, 2009.

[122] Huovila et al., 2007.

Even if market forces are put into place, consumers still might not alter their behavior. Consumers often do not recognize that they can save money by making such investments. In addition, the initial investment required for green building can be considerable, whereas the savings are not realized immediately. Thus, educational efforts on green building and cost savings are crucial.[123] Educating owners on building-envelope modifications and energy-efficient appliances can reduce the resource draw of existing buildings. Tax breaks and other financial incentives can help facilitate investments.[124]

Even with energy and water pricing, the financial incentives for reducing resource consumption might be misaligned. For example, in some renter-occupied spaces, tenants pay for monthly utilities. In these cases, owners might have little incentive to invest in efficiency technologies because their costs are fixed. Occupants might conserve resources but might have little incentive to invest in efficiency technologies because they might be short-term residents who will not remain long enough to recoup their investment.

The institute should identify prominent examples of misaligned incentives and develop policies to correct them.[125] In the example in the previous paragraph, mandated information disclosure on building-resource use could help align these incentives. If potential building users are well informed of what their utility costs will be if they decide to inhabit a building, they might be influenced toward more–resource-efficient choices and choose to pay slightly higher rents in return for lower utility costs. Such considerations are especially important in Qatar, where a large fraction of renters are short-term laborers.

White-certificate programs are another policy option. These programs set up a tradable system of permits for a specified level of consumption. Building owners are issued permits to consume resources, and those wishing to use more resources than have been allocated to them can purchase allowances from others. Such a system allows for

---

[123]Ries, Jenkins, and Wise, 2009.

[124]Ürge-Vorsatz et al., 2007.

[125]There are a few ongoing experiments with energy-use disclosure methods in Europe (Ries, Jenkins, and Wise, 2009).

efficient resource-use reduction in that cuts will occur where it is cheapest to do so. These certificate markets are addressed in the discussion of topic 14, on crosscutting environmental research.

Education Policies. The institute should also assess the impact of institutionalizing green building practices through the education of architects and engineers. The Qatar Green Building Council and Qatar University are collaborating on green building education. Such educational approaches can be effective if the building technologies are not expensive and additional incentives are not required. For example, institutionalizing the installation of intelligent and timed controls to minimize lighting and air-conditioning use might reduce energy demand without a high economic or behavioral cost.

Recommended Research: Green Buildings and Urban Planning

The green building discussion in the previous section focuses on how individual buildings can reduce resource demand. However, the institute can think more broadly about the built environment by considering how urban planning and design can be integrated with green building practices. Sustainable urban planning or smart growth considers how to design cities so that they can continue to develop and grow while minimizing the demand for resources, such as land, energy, and water.[126] Smart growth is a key concern for Qatar as cities quickly expand, resulting in urban sprawl and problematic traffic congestion.[127] The institute can help Qatar adapt by considering the central aspects to smart growth: denser cities and low-energy transportation options.[128]

Closely placed, multifamily dwellings reduce the demand for land and energy. Densely planned communities can reduce vehicle-miles traveled and accompanying fuel use by 20 to 40 percent. Energy use for climate control is also lower in these communities: Household energy consumption in the United States for multifamily dwellings is half

---

[126]EPA, 2011.

[127]Sambidge, 2009.

[128]W. Rees, 2009.

what is it for single-family homes, in part because shared walls reduce heat transfer with the outside environment.[129]

In addition to decreasing vehicle-miles through denser cities, urban design can also encourage less energy-intensive transportation alternatives, such as walking, cycling, and public transit. Design elements to increase these forms of transportation include a well-developed public transit system with safe and comfortable waiting spaces, mixed-use neighborhoods, medium to short block lengths, and safe and wide sidewalks.[130]

Thus, the institute should work with the Urban Planning and Development Authority to develop land-use legislation and zoning by-laws to encourage the smart-growth elements described earlier. First, the institute should survey the literature on smart growth, generating a preliminary set of recommendations by adapting existing case studies to suit the existing environment and culture of Qatar. As with the earlier green building discussion, public buy-in for any change is crucial, especially when behavioral change is involved. Market research on smart growth suggests that smart growth can be popular,[131] and community involvement in urban planning is known to improve the decisionmaking process and increase public satisfaction.[132] Thus, the institute should make recommendations on how to encourage community involvement in the planning process.

Recommended Research: The Institute as a Model and Test Bed for Green Building

Finally, the institute should itself serve as a model for green building design and innovation in Qatar and in the region and use its facilities as a test bed for new practices and technologies. This would demonstrate to Qatar and to the region that the institute is committed to its

---

[129] Note that there is likely a selection effect here due to a relationship between housing type, socioeconomic status, and energy use. Regarding the role of shared walls in reducing heat transfer, see Jonathan Rose Companies, 2011.

[130] Ewing, 1999.

[131] Logan, Siejka, and Kannan, undated.

[132] Healey, 1998.

mission of sustainability and that green buildings are a practical and valuable approach to achieving sustainability. The institute's campus would also provide a living laboratory for researchers at the institute to conduct technology and behavioral research.

The National Renewable Energy Laboratory (NREL) in the United States has taken this approach and is recognized as a leader in green buildings. NREL's 2009 annual report notes,

> As the nation's preeminent institution for renewable energy and energy efficiency [research and development], it is NREL's goal to provide leadership and serve as a global model for sustainability under [U.S. Department of Energy] Order 430.2B which establishes aggressive goals as it relates to energy efficiency, renewable energy use, sustainable building requirements, and conservation efforts at federal agencies.[133]

Consistent with this mission, NREL uses 100 percent renewable electricity primarily from on-site renewable energy projects and the purchase of renewable energy credits (RECs), in comparison to the 3-percent mandate set by the government for that year.[134] It has also reduced its water consumption by 27 percent, compared with the 6 percent required; has reduced its waste by 57 percent since 2003; and tracks and manages it own GHG emissions through a variety of measures.[135] NREL also uses its facilities to research, develop, and demonstrate new technologies.

Collaboration Opportunities and Human-Capital Needs for Qatar

Other organizations within Qatar have already begun exploring green building strategies. BQDRI is in the process of implementing the Qatar Sustainability Assessment System, a rating program that encourages the development of sustainable buildings through standards based on international best practices that have been modified to suit the cir-

---

[133]NREL, 2010c, p. 4.

[134]NREL, 2009, p. 14.

[135]NREL, 2009, pp. 9–14.

cumstances in Qatar. As part of the implementation strategy, the Qatar Green Building Council has partnered with Qatar University to educate building professionals, as well as school children, on green building practices. The institute could collaborate with the Qatar Green Building Council, Dohaland, GreenGulf, BQDRI, and other local organizations on data collection, best practices, and knowledge transfer.

Additionally, through collaboration with the GCC, Qatar could research how to adapt current green building design practices and codes to the local climate and use patterns. Most GCC countries already have active green building research programs, which are spurred by their growing economies and the large construction boom. SQU in Oman is arguably the largest contributor to green building research in the region. In the UAE, UAEU is conducting green building research, and the carbon-neutral city of Masdar serves as a living research lab. Kuwait is also heavily involved in green building research at KISR and at Kuwait University. Other institutions undertaking green building research include the University of Bahrain in Bahrain and KACST, KFUPM, and King Saud University in Saudi Arabia. All of these institutions represent potential collaboration opportunities for Qatar.

To best capitalize on research and implementation opportunities in the green building sector, Qatar will need a supply of engineers, architects, and city planners that are well versed in existing technologies and can work together to tailor them for Qatar. Behavioral scientists, economists, and public-policy experts will be required to determine which technologies and policies are most likely to be accepted; they can work in collaboration with policymakers and ministries to explore implementation options. Some of this expertise can be found or developed within the GCC countries. QU offers undergraduate education in urban planning and design, civil engineering, and architecture; however, KFUPM in KSA is the only university in the region to offer an accredited advanced-degree program (M.Sc.) in city and regional planning. SQU offers a Ph.D. program in civil engineering. In addition, if Qatar wishes to implement green building technologies and maintain building-performance standards on a broad scale, it will require a large pool of trained professionals; maintaining an adequate supply of qualified inspectors has been problematic in other regions.[136]

Finally, green buildings are part of the linked system between energy, water, and material producers and their consumers. Therefore, green building research and deployment incentives can be integrated with other research areas of interest, such as water demand management (topic 11) and smart grids (topic 7). As an integrated system, resource use and GHGs in Qatar could be significantly reduced.

Priority Topic 7: Smart Grids

Background and Motivation

Given its abundant energy resources, energy conservation has historically not been a priority in Qatar. However, various forces have made energy conservation an increasing concern in recent years. A rapidly growing population is increasing electricity demand by 7 percent per year.[137] Residents of Qatar also consume high amounts of electricity per capita, in part because it comes at no (or low) cost.[138] Qatar is seeking to balance its growing near-term energy needs and natural-resource consumption with the needs of future generations.[139] Simultaneously, there is increasing global pressure to reduce fossil fuel dependence and GHG emissions. Smart grid technologies provide an opportunity to reduce electricity demand. The term *smart grids* refers to a variety of flow-control and monitoring technologies that can be applied to the electrical grid to improve the reliability, security, efficiency, and flexibility of the electricity transmission and distribution system.

---

[136]Ürge-Vorsatz et al., 2007; Ries, Jenkins, and Wise, 2009.

[137]Ürge-Vorsatz et al., 2007; Ries, Jenkins, and Wise, 2009.

[138]Pearce, 2010.

[139]Energy Information Administration, undated. Qatar's per capita electricity use ranked in the top five countries globally in 2005 (International Energy Agency, 2007). Qatar is also in the top third of countries in terms of per capita residential energy consumption (International Energy Agency, 2007).

Smart grids offer several advantages for Qatar. First, they might allow Qatar to more efficiently use its natural gas supply to generate electricity by increasing the efficiency and stability of electricity transmission and distribution. Second, by providing information to consumers, policymakers, and utilities on electricity demand, they enable a wide range of demand-management policies and strategies. Third, smart grids can help Qatar integrate renewable resources to displace fossil fuels in electricity generation.

Smart grid technologies are attracting increasing interest and investment in Qatar.[140] We recommend that the institute undertake smart grid research if the benefits of smart grids—efficiency, demand management, and the use of renewables—become key energy goals for Qatar.

The institute can improve on the current smart grid efforts in Qatar through research on the following three subtopics:

- improving efficiency through grid monitoring and control
- managing demand with monitoring and incentives for efficient electricity consumption
- integrating renewables in a smart grid.

To develop a prioritized smart grid research plan, the institute should also conduct preliminary work to evaluate the future directions of Qatar's electricity generation and pricing strategies and move forward with research that is most applicable.

Recommended Research: Grid Monitoring and Control

Increased monitoring and control capacity on the grid can help Qatar improve its electricity efficiency and reliability without requiring changes in supply and end-user demand. In the near term, monitoring and control technologies can be used to detect system failures and reroute electricity to avoid them. Over a longer time period, these devices also produce usage and flow data that can be used to understand and improve grid operation. For example, data collected from

---

[140] General Secretariat for Development Planning, 2009.

monitoring devices could reveal opportunities for more directly routing electricity and thus reducing transmission losses. Such monitoring systems are being increasingly deployed.[141]

We recommend that the institute research new monitoring and routing technologies to help capitalize on these opportunities. The institute should work closely with grid managers in Qatar to assess current best practices and technologies for electrical-grid monitoring and determine how they should be adapted for the Qatari grid. In addition to technology research, the institute should conduct policy research to determine whether the benefits in electricity savings and reliability would merit the investment cost of these technologies.

In the longer term, the institute could collaborate with grid managers to study use and flow data and, for example, redesign electricity flow routes to minimize the distance traveled and thus reduce the losses incurred in transmission. Flow can also be monitored and controlled to optimize voltage and minimize losses.[142]

Even with efficiency gains such as these, however, system losses during transmission and distribution can be considerable, on the order of 6 percent or higher.[143] The reasons for these system losses are poorly understood to date, but monitoring and control technologies could help assess and address these losses. Thus, the institute could also collaborate with grid managers to better understand transmission losses. The knowledge gained, as well as any technology developed to help mitigate these losses, would be widely applicable.

Recommended Research: Managing Demand with Smart Grids
The institute should also research pricing, automation, and communication tools for consumers in order to enable demand-side management. For example, advanced metering systems show users their own

---

[141] Such countries as Japan, Australia, and South Korea are setting up commercial-scale smart grids (Hosaka, 2010). Smart grid components are currently in place in some locations, including in Italy and in the U.S. states of Colorado and Texas.

[142] Electric Power Research Institute, 2008.

[143] Electric Power Research Institute, 2008.

electricity consumption in greater detail and in real or near-real time.[144] This makes it possible to have a pricing strategy vary based on supply and demand—e.g., higher rates for electricity during peak periods and lower rates during off-peak periods. In turn, consumers can use automated appliances that run at off-peak hours and thus save electricity and money.

The institute should also research ways of using smart-meter data to better implement pricing strategies. Smart meters can help the institute understand consumer behavior in Qatar and thus help design a pricing system that will incentivize more-efficient electricity usage. Additionally, because smart meters can provide real-time information on electricity use to the consumer, the consumer can quickly modify behavior in response to incentives. For example, one pricing policy is to set a threshold on electricity use for households; rates are low below this threshold but increase once the threshold has been exceeded. Smart meters could allow consumers to more efficiently respond to such a strategy. By better understanding when and how they use electricity, consumers could eliminate unnecessary electricity use—e.g., turning off lights during the day—and bring their use below the threshold.

Because the usefulness of a smart meter lies in its ability to communicate information on usage and cost, if smart meters are adopted, the institute should also help develop educational programs to inform consumers about to use smart-meter information most effectively. In designing this program, the institute should consider that such technology might meet with initial resistance, as it has in other countries, such as the United States. The institute should aim to develop a campaign that can explain the benefits to consumers.

Other opportunities exist for demand management with smart meters that do not involve daily decisions for participation. These include automated load shedding of selected power loads during periods of high demand, with a range of incentives and strategies that can be employed to implement these load shifts.

---

[144]Litos Strategic Communication, undated, 2008.

Recommended Research: Integrating Renewables

If solar energy development becomes a priority for Qatar, the institute should conduct research on smart grid technologies that help integrate renewables into the grid. Solar electricity is a complex energy source because it is intermittent and because its generation is often not centralized.[145] The institute should work in collaboration with grid managers to develop technologies and procedures that help ensure that supply can meet demand and that minimize wasted supply.

Ensuring That Electricity Supply Meets Demand. Solar energy is unavailable at night and might not provide a steady supply even during the day. To meet supply at large scales of deployment, solar energy can be coupled with fossil fuel generation. However, if demand and supply are not closely monitored, the coupling of multiple electricity sources can result in oversupply and wastage because electricity cannot be stored on the grid. In the case of undersupply, outages will result.

Matching supply with demand becomes even more complex when electricity generation is dispersed. For example, PV can be installed on the rooftops of many individual office buildings. However, there might be a mismatch of supply and demand at each building: On cloudy days, demand might exceed the electricity produced on site; on bright days, the building might produce more than it consumes. Similarly, demand could vary on working days versus holidays and daytime versus nighttime. Thus, businesses might act as both suppliers and consumers of electricity, supplying net electricity when there is excess and consuming when there is a shortage.

At such sites, two-way meters will be required to track both supply and demand at these sites. Such technology is currently in use. However, large amounts of intermittent, dispersed electricity generation can make matching supply with demand very complex.

Thus, the institute should research applications of smart grid technologies that can more closely monitor and quickly control for changes in supply and demand. The institute should also research the impact of large amounts of intermittent, dispersed electricity generation and how smart grid technologies can balance supply and demand

---

[145] Ipakchi and Albuyeh, 2009.

in cases in which there are many suppliers. Any insight gained will be highly relevant to other countries that aim to increase generation from renewables.

Minimizing Waste. The institute should also undertake research that helps reduce waste. This includes coupling renewables integration and pricing research to determine how smart meters, incentives, and automation can be tailored to maximize the use of these intermittent electricity supplies and not waste excess at peak supply times. For example, electricity could be cheaper when the sun is shining brightly. Automated appliances that monitor dynamic prices might defer use until midday, when the sun is brightest and electricity might be cheapest. Such research by the institute would be informative for other countries in the region, as well as for any other country with substantial, intermittent renewable energy resources.[146]

The institute should also research methods to couple smart grids with electricity-storage options, which would help address intermittency problems. The institute should begin by assessing current best practices for storage. However, this is an emerging research area, and it is likely to become increasingly relevant as intermittent renewable generation is scaled up globally. Thus, the institute should also consider research to develop improved storage technology on the grid.

Collaboration Opportunities and Human-Capital Needs for Qatar

The benefits of smart grids are numerous, and they are potentially applicable to almost any electricity grid in the world. If Qatar is able to advance the leading edge of smart grid technology and policy, the market for these advancements could be large.

For these reasons, smart grids are an active area of research around the world and present many opportunities for collaboration for the institute. Among the GCC countries, the UAE has a high-performance

---

[146]Smart meters can be used to optimize the timing of electricity use. Elsewhere, smart meters are used to reduce peak demand (Litos Strategic Communication, 2008). This is desirable because, to fully satisfy the peak demand, suppliers are often forced to use less efficient electricity-generating plants or dirtier fuels. In the case of Qatar, however, peak demand is likely to be satisfied by natural gas, which is relatively clean. In Qatar, timing demand could instead help consumers use renewable energy efficiently.

grid computing laboratory that hosts the fastest super-grid computer in the region. King Saud University and KISR are also involved in smart grid research in the areas of renewable integration and demand-side management, respectively. Increased regionalization of the grid within the GCC provides an additional incentive for collaboration. Within Qatar, Texas A&M University at Qatar has several smart grid projects, and Qatar University and GreenGulf are collaborating with Chevron Energy Solutions on smart grid research.

Smart grids are also synergistic with other research that the institute could undertake, such as solar energy and green building design research. For example, if solar PV advances sufficiently, it could become an integral part of green building design in the region, and smart grids can help integrate distributed renewable energy production into the electricity grid.

Improving and implementing smart grids would require the expertise of computer scientists and electrical engineers. Economic and policy research would also help assess demand-management policies and improve design objectives. Qatar could rely on some of the expertise already resident in Education City's universities with Carnegie Mellon's computer-science program and the electrical and computer engineering program at Texas A&M University at Qatar. Additionally, there is policymaking expertise at the General Secretariat for Development and other government agencies and think tanks throughout Qatar. UAEU launched a Ph.D. program in 2009 for information technology and recently launched a Ph.D. in social sciences and humanities, which offers a concentration in geography and urban planning. KAUST in Saudi Arabia also offers a Ph.D. in computer science. These programs offer opportunities within the region to develop computer-science research expertise at a postgraduate level.

Priority Topic 8: Energy Strategy Development

Background and Motivation

Today, Qatar has opportunities to improve and diversify its energy sector. Within its oil and gas industries, there are opportunities for enhancing energy efficiency. For example, although electricity production has quadrupled since 1990, transmission and distribution losses have also increased, from 5 percent to 9 percent.[147] Qatar also has the potential to cultivate alternative, renewable energy sources. Adding renewable sources to its energy portfolio could help to supplement Qatar's future energy use as oil and gas reserves are depleted, and could also decrease Qatar's CO_2 emissions, which are currently the highest in the world on a per capita basis.[148]

Enhancements to the energy sector could occur through various channels, including an increase in efficiency of conventional energy production, or the development of one or more alternative energy sources. Importantly, activities in the energy sector are linked and affect each other. For example, if Qatar pursues distributed solar PV as a long-term renewable energy source, then smart grids could be an important way to increase the benefits of solar energy. Smart grids might not be appropriate otherwise. If Qatar seeks to maximize its exports of natural gas and petroleum, then it will be important to reduce domestic use of these resources, which could require investments in more-efficient end-use technology, energy-conservation campaigns, and a switch to alternative energy sources. If Qatar does not seek to reduce its domestic energy use in order to increase exports, then such measures might be important only if Qatar has a goal of reducing GHG emissions or preserving nonrenewable energy resources for the future.

Moreover, activities in the energy sector can have far-reaching implications that spill over into other sectors. For example, the development of an alternative source of energy, such as biomass from livestock manure or crop residues, could change livestock-management practices or cropping patterns. In addition, the prices charged to consumers, farmers, and industry for energy supplies, as well as the quantity of conventional and alternative energy supplies developed, will likely affect energy use and development patterns.

We recommend that the institute help Qatar develop a comprehensive, goal-driven energy strategy to help establish national energy

---

[147] General Secretariat for Development Planning, 2009.

[148] General Secretariat for Development Planning, 2009.

priorities and address the linkages between energy-sector development and other parts of the economy. Such a strategy can also help Qatar prioritize important but potentially competing research and development options, such as the other seven priority energy research topics we identified. Specifically, we recommend that the institute do the following:

1. Conduct an energy resource characterization (ERC) to identify potential changes to the management of current energy sources and to assess the technical and economic feasibility of developing new energy sources.

2. Provide guidance on developing long-term goals for energy use and security, including characterizing patterns and trends in demand, identifying a desired energy portfolio, and developing potential strategies for domestic energy consumption versus exports, as well as for present versus future extraction.

Recommended Research: Conducting an Energy Resource Characterization

The institute should begin by conducting an ERC. ERC is a process of gathering information that will enable an assessment of the technical and economic feasibility of developing, or enhancing the use of, an energy source. In the case of a well-established energy source (e.g., oil or natural gas), the institute could conduct research to identify state-of-the-art technology and best management practices and to evaluate gaps between current and state-of-the-art practices.

In the case of a recently developed energy source (e.g., solar power), ERC could involve identifying potential sites for development, estimating energy stocks and flows, and evaluating the variability of the resource. Some energy sources (e.g., biomass or wind) are not seen as major energy sources for Qatar. Nevertheless, the institute should characterize these resources as well, given that new or niche applications might become apparent. The institute should begin by assessing what is already known about each of its potential energy sources and identifying specific gaps in knowledge.[149] In this section, we briefly discuss the types of data that could be helpful in characterizing several leading renewable energy sources (organized alphabetically).

Biomass. Biomass energy can come from a variety of sources, including organic waste in landfills, crop residues, and livestock manure. To assess the potential for using biomass energy, the institute should conduct research to identify the locations of potential sources, the types and amounts of biofuel available at each source, and the distance of each source from the utility grid.[150] These data can then be used to study the feasibility and costs of converting the biofuel into energy and distributing the energy.

Geothermal Energy. Geothermal energy, which comes from heat under the earth's surface, can be used in several ways: direct use of hot, underground water; the use of deep, underground reservoirs of steam and hot water to drive turbines that produce electricity; and the use of geothermal pumps that can move hot air from buildings to the cooler ground when the weather is hot and from the warmer ground into buildings when the weather is cold.[151] The institute should evaluate the costs and potential energy savings from installing geothermal heating and cooling systems in buildings. The institute could also evaluate whether there are any potential reservoirs of steam or hot water that might be used to drive turbines; however, since geologically active areas (e.g., the Pacific Rim) generally have the highest underground temperatures,[152] this application of geothermal energy might not be particularly feasible in Qatar.

Solar Energy. Qatar has high levels of sunlight, making solar energy a potentially important addition to its energy portfolio. The first step in understanding the capacity for solar energy development is to install monitoring stations that can collect data on solar radiation

---

[149] The institute should also track trends and follow research being conducted elsewhere on the costs of electricity produced by these other sources versus conventional fuels. This information would help inform subsequent work on prioritizing investments in renewable energy.

[150] Jalalzadeh-Azar, Saur, and Lopez, 2010.

[151] Union of Concerned Scientists (UCS), 2009.

[152] UCS, 2009.

at potential sites in Qatar. The institute should conduct research to characterize both short-term and long-term solar-radiation characteristics. Short-term fluctuations in solar-radiation levels will influence the selection of appropriate technology, as well as the amount of storage capacity needed. A long-term assessment can be used to identify the most-appropriate sites for solar collection.[153] The data on solar radiation can then be used to estimate the amount of solar energy that could be generated from a particular installation and to conduct a technical and cost feasibility analysis.

Tidal Energy. The use of tidal energy is relatively rare, but tidal energy is more predictable than wind energy, and there is a significant amount of current research in this area. Tidal-energy devices can generate power similarly to the way wind turbines do: The kinetic energy of the periodic tidal motion is converted into mechanical or electrical energy. However, the currents around Qatar are generally not very strong (usually with speeds of less than 1 knot), and the Gulf of Salwa is protected from tidal movements,[154] so tidal energy might not be practical for Qatar. To explore the possibility of using tidal energy, the institute could measure the current produced by the water column at potential sites—for example, by using an acoustic device.[155] Data on available currents could then be used to evaluate the technical feasibility and cost of various types of turbines to harness tidal energy.

Wind Energy. Wind energy can be harnessed by using a turbine to convert the kinetic energy of the wind into electrical or mechanical energy. To understand the potential for using wind turbines, the institute should conduct research to determine the prevalent meteorology of the proposed sites, including the predominant direction and strength of the wind. The variability of wind patterns, both diurnally and across seasons, should also be assessed. The data on wind patterns

---

[153] U.S. Department of Energy, 2010.

[154] General Secretariat for Development Planning, 2009.

[155] Eppler, 2010.

could then be fed into an atmospheric model that can evaluate wind-generation potential at various sites.[156]

Recommended Research: Providing Guidance on Goals

The institute can use the results of the initial energy characterization to recommend specific areas in which conventional energy sources could be enhanced and to identify promising new energy sources for development. The institute could also provide guidance in setting and achieving near- and long-term goals.

This research should include developing estimates of Qatar's current levels and trends in energy demand, as well as the ways in which energy is used. Understanding energy demand patterns is a necessary part of an overall energy strategy. Energy use is likely to rise in line with continued population and economic growth, so the energy strategy should take into account projected future demand. In addition, the specific uses and locations of energy could affect the optimal energy portfolio: For example, if most energy is used by households and industries in Doha, then developing energy sources relatively close to Doha could help to minimize losses during transmission.

Additionally, the institute should conduct research on designing and implementing electricity pricing and other demand-management approaches to reduce domestic energy consumption and electricity waste. The institute should evaluate the effectiveness of electricity pricing, assessing the economic and social consequences and the effects on consumption. Research on electricity consumer behavior is important and is being undertaken elsewhere in the world.[157] However, local behavior in Qatar is likely to be distinct, particularly because electricity use in Qatar is free or highly subsidized. By assessing consumption data, the institute would better understand how and when electricity is being used and be able to better evaluate policies to encourage energy efficiency. A well-designed pricing system, particularly when coupled

---

[156] EnerNex Corporation and WindLogics Inc., 2004.

[157] Gridwise Alliance, 2009.

with smart grids (topic 7), could decrease demand with minimal costs to consumers.[158]

A second area of research could include developing a desired energy portfolio. For instance, based on long-term prospects for oil and gas management, as well as the feasibility of using solar energy, it might be desirable to meet a certain percentage of domestic energy demand using renewable sources (known as a renewable portfolio standard). Alternatively, a carbon portfolio standard, in which an industry or sector is required to meet a specific average GHG intensity per unit of product delivered, could be used. For example, a carbon-emission portfolio standard might require that the electricity sector reduce GHG intensity by 50 percent. The institute's initial energy characterization will be valuable in helping the government identify components of a feasible energy portfolio.

Third, the institute could conduct research on the optimal strategy for balancing domestic versus export energy supply, as well as current versus future extraction rates. The institute could undertake research on the optimal mix of natural gas exports versus domestic consumption.[159] Similarly, since oil and gas are nonrenewable resources, they will eventually be depleted; the institute can conduct research on the optimal strategy for managing the extraction of these resources over time.

The specific strategies that are appropriate for balancing domestic use versus export, as well as current versus future use, will depend on Qatar's goals and other factors, such as Qatar's export agreements, projected growth in domestic demand, and the feasibility of developing

---

[158] Smart grids can play an important role in managing demand. Smart meters can help the institute understand consumer behavior in Qatar and thus help design a pricing system that will incentivize more-efficient electricity usage. Smart grid technology can also enable flexible pricing systems. These are addressed in the discussion of topic 7, smart grids. If the institute undertakes both smart grid research and strategic energy planning research, demand management is a key topic on which the two programs should collaborate.

[159] One strategy for determining the amount to export versus use domestically is to allow domestic prices to reflect the opportunity cost of forgone exports. However, since electricity is provided for free to Qatari residents (General Secretariat for Development Planning, 2009), such market-based signals are lacking, and there is little incentive for domestic consumers to conserve electricity (and thereby natural gas).

renewable energy sources. For example, Qatar could aim to maximize profits from the extraction and use of natural gas and oil; alternatively, it could aim to ensure that sufficient oil and gas resources are available to satisfy domestic demand until renewable resources can be expected to take their place. The institute could adapt existing frameworks for optimizing resource use over time to provide guidance on how to meet Qatar's goals.[160]

Collaboration Opportunities and Human-Capital Needs for Qatar

Any national strategy requires the concurrence of different stakeholders in order to be successful. Moreover, involving stakeholders from a variety of sectors is important to ensure that the links between the energy sector and other sectors of the economy are identified and addressed. Several major stakeholder groups can be involved in this process. Many government ministries and industries, including the Ministry of Energy and Industry and the Ministry of Electricity and Water, have a direct interest in long-term energy planning. So do firms, such as Qatar Petroleum and Qatargas. Planning and research institutions can also be valuable contributors to the management and research components of the energy planning process. One key institution is the General Secretariat for Development Planning, which works with government agencies, industry, the community, and international organizations to create a national development strategy. Qatar University has also established the Gas Processing Center, which develops applied research capabilities for asset management, process optimization, and sustainable development in Qatar's gas industry, as well as an Environmental Studies Center, which seeks to promote environmental research and public awareness.[161]

In addition, opportunities for collaboration with institutions outside Qatar exist. The institute could collaborate with private or govern-

---

[160]The classic economic framework for optimizing the management of nonrenewable resources over time is reviewed by Krautkraemer, 1998. Alternative decisionmaking frameworks exist; for example, Ben-Haim, 2006, and Lempert et al., 2006, both develop decisionmaking methods to identify strategies that are robust to multiple potential future scenarios.

[161] Qatar University, 2011a.

ment agencies that conduct renewable energy research, such as NREL, the European Renewable Energy Council, or the European Marine Energy Centre (EMEC).[162] Qatar could benefit from conducting joint assessments with neighboring countries, such as Bahrain, Kuwait, or the UAE, that share similar resource characteristics, particularly for potential energy sources (such as tidal energy) that are based on shared resources.

To undertake this research, the institute would require a multidisciplinary team of researchers, including engineers and policy analysts, with experience in energy-related issues. Qatar should draw on some of its resident expertise at Qatar University, Texas A&M University at Qatar, and Carnegie Mellon University for engineering expertise and can look toward such agencies as the United Nation's General Secretariat for Development Planning for policy experts. The Masdar Institute of Science and Technology and UAEU offer local opportunities to recruit and develop research talent at the postgraduate level in multidisciplinary Ph.D. programs that focus in environmental and energy engineering disciplines. In addition, KACST in KSA recently went through an exercise to define a vision for the kingdom's energy portfolio and could be a source of research expertise and provide an opportunity for collaboration with Qatar.

---

[162] See NREL, 2011; European Renewable Energy Council, undated; and EMEC, undated.

CHAPTER FIVE

Priority Water Research

Overview of Priority Research Topics in Water

Water is a critical resource, and water security is a pressing issue in Qatar. Qatar has no permanent surface water and receives only 80 mm of rainfall on average annually.[1] Qatar's water withdrawals come from groundwater (49 percent), desalinated water (41 percent), and treated wastewater (10 percent).[2] Desalination is one of the most energy-intensive and costly sources of water.

Despite water scarcity, Qatar consumes significant amounts of water for municipal use; as a key input into critical industries, such as oil, natural gas, and electricity production; and in agriculture. Qatar's per capita water consumption is reportedly one of the highest in the world.[3] Moreover, although Qatar's desalination capabilities have tripled in the past 15 years, the amount of desalinated water produced per capita has declined due to increasing population and higher per capita consumption.[4] Qatar's groundwater is also being extracted at four times the recharge rate.[5] As a result, both the amount and quality of groundwater in Qatar are declining.

---

[1] Food and Agriculture Organization of the United Nations, 2008.

[2] Food and Agriculture Organization of the United Nations, 2008.

[3] "Expert," 2009.

[4] General Secretariat for Development Planning, 2009.

[5] Aquaterra Environmental Solutions, 2002; Supreme Council for the Environment and Natural Resources, 2002.

Climate change is likely to make water security in Qatar (and around the world) still more challenging. Climate projections suggest that the average temperature in Qatar will increase, while the average rainfall will decrease.[6] This suggests that demand will grow further while natural supply will decline.

We recommend that the institute help address water security by undertaking research in four key areas:

- improving water supplies, through desalination (topic 9) and groundwater sustainability (topic 10)
- managing water demand (topic 11)
- developing a holistic process of achieving water security through integrated water resource management (topic 12).

Priority Topic 9: Desalination

Background and Motivation

Each year, Qatar uses desalination to produce 180 billion liters of water. This water is used almost exclusively to meet municipal water needs.[7] Like those in many countries in the GCC, most of Qatar's desalination plants use thermal desalination, in which heat is used to distill fresh-

---

[6]   All projections of annual temperature for a 50-km grid cell surrounding the Doha vicinity in 2050 from 16 general circulation models and three *Special Report on Emissions Scenarios* scenarios (B1, A1B, and A2) suggest that temperatures will increase in Qatar by 2050 (see Nakicenovic and Swart, 2000). The average increase across all models and scenarios is 2.23 degrees Celsius. In 35 of the 48 cases (16 models in three scenarios), precipitation is expected to decrease, while, in the other 13, it is expected to increase. The average rainfall change across all models and scenarios (including those in which it is expected to increase) is −9.54 percent.

Data derived from the World Climate Research Programme's (WCRP's) Coupled Model Intercomparison Project phase 3 multimodel data set (Meehl et al., 2007). These data were downscaled as described by Maurer, Adam, and Wood, 2009, using the bias-correction and spatial-downscaling method (Wood et al., 2004) to a 0.5-degree (50-km) grid, based on the 1950–1999 gridded observations of Adam and Lettenmaier, 2003.

[7]   Food and Agriculture Organization of the United Nations, 2009.

water from salt water.[8] Qatar specifically uses multistage flash (MSF), in which seawater passes through a series of chambers in which part of the freshwater component is vaporized then is condensed. MSF is a reliable and mature technology and can accommodate the high concentrations of salt and organic compounds found in the region's seawater. MSF plants can also have high capacity.

The major drawback of MSF and other thermal desalination plants is that they are energy-intensive and typically powered by fossil fuels. The estimated heat required by thermal desalination ranges from 145 to 390 kilojoules (kj) per kilogram. In other words, desalinating 1,000 liters of water requires as much energy as driving a car approximately 25 to 65 miles. As we discuss later, Qatar and other countries in the region have addressed this requirement in part by colocating thermal desalination plants with thermal power plants. This allows waste heat from the power plant to be used by the thermal desalination process, thus increasing efficiency. Nevertheless, the energy requirements even with these gains remain high.

Though Qatar has made substantial investments in desalination capacity in the past two decades, per capita production of desalinated water is falling as the population expands rapidly and the Qatar economy grows.[9] Additionally, the Qatar National Food Security Programme (QNFSP) seeks in part to increase domestic agricultural production and to use desalination to meet its water needs.[10] Thus, there is a need for Qatar to expand its desalination capabilities. Qatar must decrease the fossil fuel energy requirements from desalination in order to meet sustainability goals, such as reducing CO_2 emissions. Additionally, as Qatar's oil and natural gas resources are depleted over time,

---

[8]   General Secretariat for Development Planning, 2009.

[9]   General Secretariat for Development Planning, 2009.

[10] According to interviews with members of the QNFSP, Qatar imports approximately 90 percent of its food. Agriculture represents a negligible portion of the gross domestic product (Food and Agriculture Organization of the United Nations, 2008). The QNFSP is aimed at increasing domestic agricultural production and strengthening the security of food imports to diminish food supply deficit. Moreover, according to its mission, the QNFSP intends to use desalinated water instead of groundwater in order to prevent further degradation of groundwater. The QNFSP's mission statement can be found at QNFSP, undated.

and if natural gas becomes more valuable as an export commodity because of carbon constraints, the opportunity cost of using Qatar's oil resources to power desalination will rise.

In order to efficiently maintain or increase desalination capacity, the institute should undertake research into technologies that can improve the energy efficiency and reduce the cost of desalination. In this section, we discuss five specific areas in which research could be particularly promising for Qatar:

- Evaluate and develop hybrid thermal/reverse osmosis (RO) desalination systems.
- Assess the potential for using solar energy to supplement fuel used in thermal desalination or electricity used in membrane desalination.
- Explore cutting-edge, alternative desalination methods, such as freeze desalination and dewvaporation.[11]
- Investigate improvements to materials and processes used in existing thermal desalination plants.
- Identify future sites for colocated power and desalination plants, and assess potential for improving and expanding use of waste heat or low-grade heat from colocated power plants in the desalination process.

Research on non–energy-related environmental impacts of desalination plants, such as damage to marine life from water intake and brine disposal, is discussed in the environmental characterization section (topic 13).

Recommended Research: Hybrid Thermal/Reverse Osmosis Desalination Systems

MSF is attractive for Qatar and other countries where traditional energy sources are plentiful. Nevertheless, alternatives, such as membrane desalination (the other main form of desalination), could have

---

[11] Dewvaporation is the process of humidifying a stream of air with saline water, heating the air, then condensing the vapor to produce desalinated water.

some key advantages. The dominant type of membrane desalination is RO, in which water is pumped through a membrane under high pressure, leaving a concentrated salt solution behind. A major advantage of RO is that the energy needed to desalinate one unit of water is between one-third and one-sixth of the energy required by MSF.[12] However, although all desalination processes require some pretreatment of the entering water, thermal systems generally require only modifying the source water to avoid the scaling or corrosion of pipes, while RO systems also require the removal of mineral or biological materials to avoid membrane fouling.[13]

The institute should conduct research to determine the costs and benefits of using RO versus MSF for future power plants. Importantly, hybrid MSF/RO systems can provide the advantages of both systems and might be applicable to both existing and future plants. Many existing hybrid MSF/RO systems currently operate the MSF and RO desalination processes in parallel, which offers some advantages. For example, thermal desalination plants in Qatar are typically colocated with power plants to take advantage of waste heat. If a combined MSF/RO plant is colocated with a power plant (as is the case for the recently built Fujairah desalination plant in the UAE), overall efficiency could be further enhanced. During periods of high electricity demand, the power plant will produce a large amount of waste heat, which can be used as an input to the MSF process. During periods of low power demand, less waste heat will be produced by the power plant, thereby increasing the supplemental heat (nonwaste heat) required for the MSF process. If the RO unit is brought online during periods of low power demand, it can use electricity from the power plant and offset some portion of MSF desalination.[14] MSF also offers benefits for the RO process. RO membranes typically must be replaced every few years due

---

[12] Kalogirou, 2005.

[13] National Research Council, 2008. Specific pretreatment depends on the source water and can be accomplished through various physical or chemical methods (e.g., coagulation and sedimentation to remove solids, addition of chlorine to control biological matter, addition of sodium bisulfite to prevent corrosion) (National Research Council, 2008).

[14] Al-Mutaz, 2005; National Research Council, 2008.

to fouling. If outputs from the MSF and RO components are blended, the RO component can be operated to produce permeate (water that passes through the membrane) with relatively high concentrations of remaining salt, thus prolonging the life of the membranes.[15]

However, to take full advantage of the hybrid system, the two processes could be more fully integrated, for example, by using the pre-heated water from the MSF process to increase efficiency in the RO process. Additionally, a tightly integrated hybrid system can recover more freshwater from seawater by running the water through two desalination processes instead of one.[16] This type of integrated process is still in its infancy, though some MSF/RO hybrid plants have been tested in experimental settings.[17] Therefore, the institute should research tightly coupled hybrid systems. In addition, since most of Qatar's desalination facilities use MSF, the institute should research the feasibility of integrating an RO process into the existing MSF facilities.

Recommended Research: Solar Desalination
Another promising avenue for reducing the use of nonrenewable energy in desalination is to supplement conventional fuels with solar power. Solar-driven desalination can be either direct or indirect. Direct solar desalination uses energy from the sun to directly evaporate and condense seawater, thus producing freshwater condensate. Given the relatively low productivity of this method, it is generally considered appropriate for small-scale applications.[18] Indirect solar desalination, on the other hand, couples the collection and storage of solar energy with conventional desalination processes. The most-promising combinations are (1) a PV solar energy collection system with RO and (2) a solar thermal collection system with MSF or multieffect distillation (MED). These

---

[15]   Hamed et al., 2000.

[16]   Hamed et al., 2000.

[17]   Gude, Nirmalakhandan, and Deng, 2010.

[18]   Qiblawey and Banat, 2008.

technologies are in the pilot plant stage, with small installations in various countries, including Kuwait, the UAE, Spain, Italy, and Japan.[19]

Additional flexibility and efficiency could be gained by using solar energy to provide thermal energy for an MSF desalination plant, as well as electricity, which could be used in an RO desalination system or for other power needs. A small plant in Kuwait integrates a hybrid MSF/RO system with such a dual-purpose solar collection system.[20]

Some challenges remain in using solar energy to power desalination plants. The cost of such systems remains a barrier. The per-unit cost of using solar thermal energy with an MED plant is somewhat higher than per-unit cost from conventional energy, while the per-unit cost from a combined PV/RO plant is significantly higher due to the high cost of PV.[21] However, as conventional energy sources are depleted and solar desalination technology becomes more mature, solar desalination is likely to be more cost-competitive. In addition, given the large capital costs of both solar collectors and desalination plants, the per-unit costs of solar desalination should fall as larger-scale plants are developed. The variability of power produced by solar collection systems presents another technical challenge for MSF, MED, and RO systems, though one private company has developed a modified version of the MSF process that could accommodate variable heat sources.[22] Another technical challenge is specific to the PV/RO process: The batteries used to store electricity often fail prematurely, particularly in hot climates, exhibit leaks, and have relatively low efficiency. Current research is addressing these issues in a few ways, including evaluating PV systems without batteries and coupling RO with other processes that lower energy consumption.[23]

---

[19] Kalogirou, 2005.

[20] Kalogirou, 2005.

[21] Gude, Nirmalakhandan, and Deng, 2010.

[22] Kalogirou, 2005; Qiblawey and Banat, 2008.

[23] Gude, Nirmalakhandan, and Deng, 2010.

Recommended Research: Alternative Desalination Processes

The institute should also consider research in alternative desalination methods. For example, it is possible to freeze seawater to remove salt: When water freezes to form ice crystals, salt is excluded from the crystalline structure. The major advantage of using freezing rather than evaporation is that the energy required to change water from a liquid to a solid is an order of magnitude lower than the energy required to change water from a liquid to a gas.[24] The salt can then be washed off the ice crystals and the ice melted to provide freshwater. Several pilot plants for freeze desalination have been developed, but technical challenges remain, including difficulty in washing off the salt without melting the ice, as well as complex maintenance procedures.[25]

A second area of research focuses on processes that could allow the use of low-grade or waste heat in desalination. One such process is low-temperature thermal desalination (LTTD). The basic idea behind LTTD is to pump hot, surface seawater into a low-pressure chamber, where (because of the low pressure) a portion of the freshwater evaporates using its own heat. Cold seawater from a lower depth is pumped through the condenser, thus allowing the evaporated water to condense back into freshwater. Three small LTTD plants (with capacities between 100 and 1,000 m³ per day) have been built in India.[26] Another process that might allow the use of low-grade heat is membrane distillation, in which seawater is heated to produce a vapor and the vapor is passed through a membrane that allows only freshwater vapors through. Again, research in this area is in its initial stages, and the process has not yet been commercialized.[27] Another process, dewvaporation, involves humidifying a stream of air with seawater then condensing the vapor from the air into freshwater. A small dewvaporation pilot plant has been constructed in Arizona.[28]

---

[24] National Research Council, 2008.

[25] Khawaji, Kutubkhanah, and Wie, 2008; National Research Council, 2008.

[26] Sistla et al., 2009.

[27] Chen, Wang, and Yang, 2007; National Research Council, 2008.

[28] Beckman, 2008.

Recommended Research: Improving the Materials and Processes Used in Thermal Desalination

Thermal desalination is a mature technology, but there are potential cost and energy savings from improving the materials and processes used in MSF plants. The standard materials used for heat transfer in these plants are copper, aluminum, and titanium, but recent research indicates that the use of alternative materials, such as high-density polyethylene and polypropylene (PP), in heat-transfer surfaces could decrease energy consumption.[29]

Improving the thermal desalination process by increasing the number of stages through which the seawater is processed, or by increasing the top brine temperature, can also reduce energy requirements. Cutting-edge research has considered the use of nanofiltration prior to thermal treatment to allow for higher top brine temperatures.[30]

Research into improving the materials and processes used in current thermal desalination facilities would not fully address the growing scarcity of municipal water in Qatar, but it could provide some cost and energy savings and might be more feasible in the short run than switching to alternative desalination technologies or renewable energy sources.

Recommended Research: Colocation of Desalination and Power Plants

Thermal desalination is competitive with RO in many countries, despite its large energy requirement, because desalination plants can be colocated with power plants. Cost savings can be realized from joint construction of intake and discharge locations; in addition, as discussed earlier, waste heat from the power plant can be used as an input to the thermal desalination process, which increases efficiency by 10 to 15 percent.[31] Qatar already has three integrated desalination and power plants in Ras Laffan and Mesaieed, with plans to build a fourth in 2011. Nevertheless, the institute could consider policy research to fur-

[29] Scheffler and Leao, 2008.

[30] Hamed, 2006.

[31] National Research Council, 2008.

ther improve desalination and energy plant siting and to assess opportunities for using low-grade heat in existing plants.

First, research could help determine how low-grade heat from power plants could be used. Low-grade heat is heat that is only slightly higher than ambient temperature, while waste heat is heat that is produced in the power plant but not used and can be significantly above ambient temperature. Currently, many thermal desalination plants that are colocated with power plants already use waste steam (which might or might not be low-grade heat) from the power plant's turbines in the desalination process. Several alternative desalination technologies that make use of low-grade heat were discussed earlier and might provide additional opportunities for using such heat from colocated power plants.[32]

Second, research can help to better site colocated power and desalination plants. For example, development patterns are important because transmission and distribution costs will be minimized if the plants are located close to areas where the power and water will be used. However, proximity might need to be balanced with public perceptions about the desirability and safety of having plants nearby. This concern might warrant a buffer between the plants and residential areas. Since conventional power plants and desalination facilities take in water for cooling and produce liquid waste streams, an additional consideration is to locate their intake and discharge pipes in such a way as to minimize potential damage to aquatic life. This issue is addressed in more detail in our discussion of topic 13, on environmental characterization.

Encouraging research on colocation issues is therefore important because it might provide energy savings and can also serve to identify and address some of the site-specific social and environmental challenges associated with power and desalination plants.

Collaboration Opportunities and Human-Capital Needs for Qatar

Many of the technologies discussed thus far are in their infancy, and many countries, both in the GCC and beyond, have a strong interest in reducing the energy use of their desalination plants. Given Qatar's

---

[32] National Research Council, 2008.

high levels of sunlight and long history of operating thermal desalination plants, it could take a leadership role in optimizing and scaling up solar-powered desalination plants. The key expertise needed for Qatar to pursue this research agenda is engineers in desalination (preferably with some knowledge of RO, as well as thermal desalination) and solar power generation. Some of this capacity already exists in Qatar with researchers at QU and Texas A&M University at Qatar engaged in applied solar research. There are some highly regarded postgraduate degree programs for mechanical and industrial engineering in the Gulf region, and Qatar could look regionally to attract graduates from multidisciplinary degree programs that have an alternative energy focus, such as Masdar Institute of Science and Technology in Abu Dhabi or KAUST in Saudi Arabia.

In this line of research, Qatar could collaborate with private entities or government agencies in Kuwait, the UAE, Australia, Japan, Spain, Mexico, the United States, Germany, and several other countries that are involved in pilot studies. Germany has been a global leader in solar innovation. The Bavarian Center for Applied Energy Research, the Centre for Solar Energy and Hydrogen Research, and the Ludwig-Maximilians-Universität München, as well as some small private firms, are conducting MEDRC-funded research on the use of solar energy for thermal desalination and design of hybrid solar-desalination plants.[33] Within the Middle East and North Africa region, Israel has been a leader in research on solar integration and innovation in desalination at the Jacob Blaustein Institute for Desert Research. In the GCC, there might be opportunities to collaborate with Masdar Institute, which is trying to develop as a leader in alternative energy. Masdar Institute also has access to Masdar City, a sustainable urban development, as a living laboratory for pilot studies. Additionally, Kuwait University and KISR have a portfolio of research in thermal and membrane desalination technologies. Researchers at SQU are investigating desalination optimization technologies and low-tech thermal desalination innovations for remote areas. Given international interest in using renewable

---

[33] MEDRC is based in Oman and funds desalination research globally, although entities seeking MEDRC project grants must have a partner in the MENA region.

energy to produce freshwater, if Qatar is successful in commercializing large-scale solar desalination, it would be in a strong position to take a leadership role in exporting this technology.

Priority Topic 10: Groundwater Sustainability

Background and Motivation

Qatar's only natural source of water is groundwater.[34] Two-thirds of the land surface is made up of some 850 contiguous depressions with interior drainage and with catchment areas. Direct and indirect recharge of groundwater from 50 to 80 mm of rainfall per year forms the main natural internal water resource.[35]

Qatar's groundwater is overexploited. Groundwater is being extracted in Qatar at four times the sustainable yield of the aquifer.[36] As a result, both the amount and quality of the groundwater resources are declining. There is a 3-percent annual decline in the level of groundwater reserves, and the groundwater is becoming increasingly saline as seawater and saline water from deeper underground have intruded to replace extracted water.[37] Moreover, demand for all water, including groundwater, in Qatar is likely increasing given the growing population, economic development, a desire to expand agriculture and establish food security, and no pricing system to manage demand.[38] Although Qatar has explored various methods for enhancing groundwater recharge and recovery, these efforts are not likely to counteract these factors.

---

[34] Qatar has no permanent surface water, and rainfall is minimal.

[35] United Nations, 1997; Food and Agriculture Organization of the United Nations, 2008.

[36] Aquaterra Environmental Solutions, 2002; Supreme Council for the Environment and Natural Reserves (SCENR), 2002.

[37] Aquaterra Environmental Solutions, 2002; Food and Agriculture Organization of the United Nations, 2008; United Nations, 1997; SCENR, 2002.

[38] The per capita availability of desalinated water has fallen during the past few years, despite the fact that Qatar has made substantial investments in expanding its desalination (General Secretariat for Development Planning, 2009).

The consequences of this level of exploitation are considerable. First, Qatar's groundwater might be completely depleted in the coming decades. This would leave Qatar wholly dependent on desalination and imported water, presenting a significant vulnerability in Qatar's plans for sustainable water security. Additionally, given that the majority of groundwater extracted in Qatar is used for agriculture, the use of this high-salinity water for irrigation has resulted in the deterioration of cropland and has prompted farmers to abandon land when it is no longer suitable.[39] Moreover, rates of seawater ingress might not be linear but might pivot on a threshold. Thus, once the aquifer is depleted beyond a particular point, subsequent small changes in groundwater levels might allow significant levels of saltwater intrusion.

The institute should therefore develop a comprehensive groundwater sustainability research program. This includes four key areas of research:

- monitoring groundwater
- reducing demand for groundwater
- increasing groundwater recharge through natural and artificial means
- understanding the effects that climate change could have on the long-term sustainability of groundwater resources.

Recommended Research: Monitoring Groundwater

The sustainability of Qatar's groundwater depends largely on its agricultural sector. In 2005, nearly 100 percent of Qatar's groundwater—approximately 220 m^3—was used for agriculture (irrigation and livestock). This represents about 50 percent of the total water withdrawn in the country. What we do not know, with a level of clarity that could support policies to optimally control groundwater extraction, is how groundwater use is distributed among particular aquifers, wells, and farms. In 2004, for example, 945 of 1,192 registered farms in Qatar were operative, representing 49 percent of the area actually equipped

---

[39] SCENR, 2002; Food and Agriculture Organization of the United Nations, 2008.

for irrigation.[40] This suggests that the potential for groundwater extraction could be much higher than already-unsustainable levels. Further, although some information is known about the types of irrigation being used, estimates vary about irrigation efficiency, due to a lack of systematic monitoring.

Thus, research should be undertaken to establish a system to monitor both groundwater resources and extraction. Qatar has started a water monitoring development program, involving a telemetry system at three automatic agrometeorological stations, 25 hydrometeorological stations, and 48 hydrogeological stations. These automatic stations provide data for irrigation scheduling and designing irrigation systems.[41] Further research could include more-comprehensive monitoring of groundwater extraction at the point of use, beginning with a complete survey of groundwater wells and irrigation sites.

Recommended Research: Reducing Demand for Groundwater

Although agriculture is almost entirely responsible for consuming Qatar's groundwater, Qatar imports approximately 90 percent of its food, and agriculture represents a negligible percentage of the country's gross domestic product.[42] This suggests that the way in which this resource is being used is not consistent with its value. Thus, complementary research could focus on improving the efficiency of groundwater use, particularly for irrigation, which represents the largest burden on groundwater resources. This can be done in two broad ways: reducing the demand for water services in agriculture, and reducing the use of groundwater in particular to meet those demands. We discuss these in turn.

Importantly, Qatar is currently seeking to improve its food security and has created the QNFSP with the objective of increasing domestic agricultural production and strengthening the security of

---

[40] Food and Agriculture Organization of the United Nations, 2008.

[41] Food and Agriculture Organization of the United Nations, 2008.

[42] The 90-percent figure was quoted in our interviews with Qatar National Food Security Program staff; gross domestic product information comes from Food and Agriculture Organization of the United Nations, 2008.

food imports to diminish food supply deficit. Moreover, according to its mission, the QNFSP intends to use desalinated water instead of groundwater in order to prevent further degradation of groundwater. It also seeks to optimize irrigation and agricultural operations to improve water efficiency.[43] Given that significant changes to Qatar's agricultural sector are on the horizon, research in groundwater sustainability should be undertaken with those initiatives in mind. It should also be conducted in close collaboration with institutions, such as the QNFSP, that are conducting agriculture research or making agriculture policies.

Recommended Research: Reducing Demand for Water Services

Most research into reducing water use in irrigation has focused on best agriculture management practices and improving water-use efficiency through irrigation technologies. Best management practices include, for example, selecting crops based on their water needs and suitability for the environment, preparing farms with water-containment structures, and scheduling irrigation based on soil type and weather.

One efficient irrigation technique is drip irrigation, in which a small amount of water is piped to each plant through a hose, thus targeting water where it is needed and lowering evaporation losses.[44] However, there are challenges to using drip irrigation systems (and similar systems), including high capital investment costs and technical hurdles, such as clogging of the system due to high salt content, which is present in groundwater in Qatar. In addition, it is often necessary to apply fertilizers and chemicals through irrigation systems. Ongoing research is addressing several of these issues.[45] The institute should develop efficient irrigation systems that are robust to these and other challenges in the region. The institute could also conduct agricultural research to

---

[43] The QNFSP's mission statement can be found at QNFSP, undated.

[44] In contrast, more-conventional methods sprinkle or flood an entire field with water. This water therefore does not target areas that are most in need of it, and much of the water is lost to evaporation.

[45] See Hanson, Šimůnek, and Hopmans, 2006; Puig-Bargués et al., 2010; and Roberts et al., 2008.

develop ideal crops for efficient water use, potentially including investigation of genetic modification of the most-desired crops.

Recent and current research in Qatar in this area includes determining crop water requirements of the major crops in Qatar and the optimum use of water resources in agriculture and modernizing irrigation in Qatari farms.[46] Irrigation efficiency in Qatar is approximately 45 percent, despite a 90-percent extraction efficiency from wells.[47] There is great potential for increasing water-use efficiency by shifting from surface irrigation techniques to sprinkler and localized irrigation. Modern irrigation techniques could save 35 to 40 percent of the present crop water consumption.[48]

In addition to these technical approaches, a wide range of policies can reduce water demand. Perhaps most important of these is the one that makes water free to Qatari farmers, thus creating no financial incentive for farmers to conserve water or invest in the aforementioned technologies. Extensive research on how *residential* consumers react to changes in the price of water has established this to be the case under a simple pricing system, but it is unclear how this research translates to potential *farm* irrigation control policies via pricing, including more-complex schemes, such as block pricing, in which initial units cost less than subsequent units.[49] Thus, the institute should conduct research to understand the role of pricing in managing agriculture water demands and to develop methods of implementing such schemes given the culture and status quo in the sector.

Other policy methods that the institute should examine include the following:

---

[46] Department of Agriculture and Water Research (DAWR), 2006a, 2006b.

[47] Food and Agriculture Organization of the United Nations, 2008.

[48] Food and Agriculture Organization of the United Nations, 2008.

[49] On simple pricing systems for residential consumers, see, e.g., Dalhuisen et al., 2003, and Espey, Espey, and Shaw, 1997. For more about farm irrigation policies, see Olmstead, Hanemann, and Stavins, 2007.

- Metering wells. The metering of wells establishes a mechanism to measure how much water is extracted and, importantly, provides a basis for charging for water use.
- Permitting of new well bores. This would require someone desiring a new well to apply (and perhaps pay for) a permit. Approval of such a permit could be contingent on many factors, including water quality and availability, effect on nearby wells and the aquifer, and environmental impacts. A permitting process for new wells would allow regulators to identify and manage new demands on groundwater. It also is a mechanism to impose a cost structure on groundwater extraction, if desired. A continual permitting system (one that requires the renewal of permits over time) would provide a source of monitoring data.
- Incentivizing the use of crops that are suited to water availability and quality. Different crops grow better in different water conditions. Where water abundance and quality is low, as in Qatar, growing crops best suited to those conditions would provide the greatest opportunity for conservation. However, farmers who receive water for free are incentivized to grow crops that bring the best market price, regardless of the amount of irrigation required. If farmers were charged for water use, their incentive would shift to balance water cost and crop value. If water continues to be available without cost (or at very low cost), however, farmers could be subsidized to grow crops that are less valuable at the market but require less water (or, conversely, they could be taxed on crops that require more water).

The institute should conduct policy research to assess the effectiveness, costs, and benefits of these price and nonprice policies.

Recommended Research: Other Agriculture Water Sources

An alternative approach to alleviating the overuse of groundwater for irrigation is to use water from other sources. This approach is particularly attractive for Qatar, where the high salinity content of groundwater makes its use for irrigation undesirable, resulting in the abandonment of cropland after several seasons of irrigation. There is considerable research to suggest that other sources of water are suitable for irrigation, including treated municipal wastewater (reclaimed water) and desalinated water. Optimizing the water supply and demand for irrigation would be part of a broader integrated water resource management (IWRM) research effort (topic 12).

The amount of treated municipal wastewater in Qatar in 2005 (98 percent tertiary treatment) was 55 million m^3.[50] This amount is only a fraction of the 262 million m^3 of total water withdrawal for agricultural purposes that year. However, it presents an opportunity to reduce groundwater extraction for irrigation, especially if combined with other sources—if policies and infrastructure are in place to transport it where needed and if it does not significantly offset other uses. In fact, approximately one-quarter of treated municipal wastewater in Qatar was already being piped to Doha for landscape irrigation, and the rest was supplied to two farms for forage crops, free of charge, in 2005.[51]

Experience with using reclaimed water is extensive and growing. Reclaimed municipal wastewater is used worldwide for irrigation. In addition to agricultural use, reclaimed water has been used for toilet flushing; fire hydrants; irrigation of landscaping; recharging of groundwater and surface waters, including recreational water; and industrial and construction uses.[52] Many countries have established standards for the use of reclaimed water, with public health as a primary concern.[53] For example, pathogens can be transmitted via irrigation (e.g., irrigation with poorly treated water has been shown to be a major source of enteric disease).[54] Although there is comprehensive research, and poli-

---

[50] Public Works Authority, 2005; Food and Agriculture Organization of the United Nations, 2008.

[51] DAWR, 2006b; Food and Agriculture Organization of the United Nations, 2008.

[52] Shelef, 1991; Richardson, 1998; Snyder et al., 2002; SCENR, 2002; Banks, 1991; EPA, National Risk Management Research Laboratory, and U.S. Agency for International Development, 2004; Dorica, Ramamurthy, and Elliott, 1998.

[53] National Research Council, 1996; EPA, National Risk Management Research Laboratory, and U.S. Agency for International Development, 2004; Fatta et al., 2004.

[54] Shuval et al., 1986.

cies exist for the use of reclaimed municipal wastewater for irrigation, there are other obstacles to its use that could be addressed through the institute's research, including what infrastructure would be required to deliver the water where it is needed and what the overall costs would be for its use.

Similarly, the idea of using treated industrial wastewater for irrigation has also been studied, including a study performed by RAND on the use of water produced by gas-to-liquids processes in Qatar. Such use is generally unprecedented, with only one documented case of *intentional* use of industrial wastewater for irrigation.[55] However, other cases of *unintentional* use of industrial wastewater for irrigation have been documented, and laboratory studies have been conducted on the feasibility and safety of this approach.[56] In general, these studies indicated an initial increase in crop productivity but then a decrease as the proportion of wastewater in the irrigation water increased. More research is needed to determine the feasibility of this approach, including the availability and quantity of treated industrial wastewater and a comparison of the risks and costs versus using groundwater, which is being rapidly depleted and is of poor quality.

Another potential source of water for irrigation is desalinated water, which represents a much larger volume (180 million m^3 in 2005) than groundwater use for agriculture.[57] At the time of that estimate, construction of additional desalination facilities in Qatar was continuing, so the potential amount of desalinated water is likely higher. However, although the use of this water for irrigation would offset groundwater use, it would also likely divert it from other uses, such as potable water, and it would also require transport to cropland.

The institute's research on the use of alternative sources of water for irrigation should include cost, infrastructure, and safety consider-

---

[55] Jimenez-Cisneros, 1995.

[56] Aziz, Inam, and Samiulla, 1999; Aziz, Manzar, and Inam, 1995; Rajannan and Oblisami, 1979; Sahai, Jabeen, and Saxena, 1983; Sahai, Shukla, et al., 1985; Misra and Behera, 1991; Fazeli et al., 1998.

[57] Qatar Electricity and Water Company, 2007; Food and Agriculture Organization of the United Nations, 2008.

ations, as well as the overall benefit (i.e., how much alternative water is available and what the offset to groundwater would be).

Recommended Research: Groundwater Recharge

In addition to reducing groundwater extraction and improving the efficiency of groundwater used, research also focuses on increasing the amount of groundwater available, by increasing the amount of water introduced to aquifers, a process known as recharge.

Qatar began a program in 1986 to increase the natural recharge of aquifers. The program involves the drilling of injection wells in depressions to depths that reach water-bearing formations, to accelerate the natural recharge of floodwater. Between the program's inception in 1986 and the time of a DAWR report in 2006, 341 recharge wells had been drilled.[58] The goal of the project is to make rapid recharge possible from the occasional storm runoff that accumulates in depressions, before it is lost through evaporation. Numerous methods have been used to monitor the success of such natural recharge efforts, and future research in Qatar could survey these methods to determine how its efforts are affecting groundwater resources.[59] More specifically, Qatar could consider research techniques specific to arid regions because the more arid the climate, the smaller and potentially more variable is the recharge flux into an aquifer.[60]

In addition to providing methods to increase natural groundwater recharge, injection wells can be used to prevent seawater intrusion into aquifers, increasing the quality of the groundwater.[61] This would be accomplished by introducing water and thus creating a subsurface pressure barrier against intruding seawater. With the considerations mentioned earlier about safety and the availability of water, the institute could investigate whether this is a suitable use of municipal or

---

[58] DAWR, 2006a; Food and Agriculture Organization of the United Nations, 2008.

[59] Simmers, 1987.

[60] See Allison, Gee, and Tyler, 1994, and Maréchal et al., 2006.

[61] Johnson, 2007.

industrial wastewater. Significantly more research is needed in this area.

Recommended Research: Climate Change and Groundwater Sustainability

Groundwater relies solely on rainfall for natural recharge. The potential effects of climate change are highly uncertain; climate change could result in a decrease in rainfall, exacerbating Qatar's groundwater problem. It is possible that climate change will result in more rainfall, but the level of uncertainty is sufficiently high that this possibility is not a good planning factor for increasing Qatar's groundwater sustainability. This area requires a great deal of research, including modeling possible futures in which rainfall amounts are significantly altered.

Collaboration Opportunities and Human-Capital Needs for Qatar

Most countries in the GCC and in MENA more broadly, face groundwater depletion. This is largely a product of the climate, geography, and growing populations and economies. Qatar could be a leader in groundwater sustainability in arid climates and could export this expertise to nations with similar challenges. Qatar is well positioned because of its leadership in groundwater recharge programs but could do more to manage groundwater use. There could be ample opportunity for the institute to collaborate with institutions in other countries, particularly with those that share Qatar's groundwater problems, climate, and geology. For instance, UAEU, SQU, and Kuwait University are all undertaking research to model groundwater quality, while Kuwait University is additionally researching use of nanoparticles for groundwater remediation. UAEU also has a variety of ongoing projects examining seawater intrusion in groundwater and simulations for groundwater recharge.

In order to carry out a broader research agenda in groundwater sustainability, the institute needs researchers in policy and economics, hydrogeologists, agronomists, and engineers (e.g., civil, environmental). Qatar already has some experts in these disciplines at Texas A&M, Qatar University, Georgetown University, and RQPI. Although Qatar does not offer master's or Ph.D. programs in this area, some expertise can be found or developed regionally. UAEU offers an M.Sc. in water resources; Ajman University of Science and Technology in the UAE offers a special M.Sc. program in groundwater engineering; Masdar Institute of Science and Technology offers an M.Sc. in water management and environmental engineering leading to a multidisciplinary Ph.D. in management, and SQU in Oman offers both M.Sc. and Ph.D. programs in soil and water management.

Priority Topic 11: Managing Water Demand

Background and Motivation

Qatar has made substantial investments in developing its main sources of water: desalinated seawater, groundwater, and treated wastewater. In the past 20 years, the production of desalinated water has risen from less than 100 million m^3 to more than 300 million m^3.[62] Qatar has also carried out projects to increase the recharge of groundwater, including drilling recharge wells, considering the possibility of artificial recharge mechanisms, and exploring the use of deep aquifers. Moreover, Qatar is a pioneer in the use of treated wastewater for irrigation of landscapes and forage crops; treated-wastewater use accounts for 10 percent of total water withdrawals in Qatar.[63]

However, increasing the supply of water is only one side of attaining water security. Rapid population growth means that the per capita availability of desalinated water has actually been falling in the past few years, and the rate of groundwater withdrawals is far above recharge rates.[64] Therefore, it is imperative that Qatar consider options for decreasing the demand for water as well. Currently, residential consumers have little incentive to conserve water: Qataris receive residential water for free, while non-Qataris pay less than half the cost of

---

[62] General Secretariat for Development Planning, 2009.

[63] Food and Agriculture Organization of the United Nations, 2008.

[64] General Secretariat for Development Planning, 2009; Food and Agriculture Organization of the United Nations, 2008.

delivered water.[65] Similarly, farmers receive water for free, and subsidized fuel and electricity also help to hide the full cost of pumping groundwater. This is reflected in the fact that only 11 percent of irrigated areas use water-efficient irrigation methods.[66] Simultaneously, Kahramaa, the Qatar General Electricity and Water Corporation, estimates that nearly 45 percent of water use is accounted for by losses from the transmission and distribution system.[67]

The institute should therefore undertake a comprehensive research agenda to reduce the consumption of water, which is often referred to as demand-side management (DSM). DSM encompasses a variety of techniques to reduce municipal, agricultural, and industrial water use, such as monitoring and metering water use, charging for water, encouraging the use of water-conserving fixtures, and education about conservation.

Other countries have implemented such DSM approaches. For example, the U.S. Environmental Protection Agency (EPA) compiled a list of case studies of residential DSM programs and has documented reductions in water use that generally ranged between 10 and 30 percent.[68] However, the success of any particular approach depends on local factors. For example, the effectiveness of using prices to encourage water conservation depends on how sensitive consumers are to price changes, which can vary across different seasons and consumer groups. Similarly, the potential benefits of nonprice mechanisms to reduce water use, such as the requirement for low-flow showerheads, will depend on water-use patterns that are specific to Qatar. The institute should therefore conduct research to determine which DSM policies are most likely to be effective in Qatar.

---

[65] General Secretariat for Development Planning, 2009.

[66] Food and Agriculture Organization of the United Nations, 2008.

[67] General Secretariat for Development Planning, 2009.

[68] EPA, 2002.

Since municipalities and agriculture account for 98 percent of water use in Qatar,[69] we recommend that the institute undertake research in four areas to address water use in these sectors:

- residential water DSM
  - development and evaluation of an efficient and equitable water pricing system
  - evaluation of the effectiveness of nonprice policies
- agricultural water DSM
- assessment and improvement of the water transmission and distribution system
- development of educational programs to create a culture of water conservation.

Since agricultural withdrawals are largely from groundwater sources, and essentially all of Qatar's groundwater withdrawals are used in agriculture, we address agricultural DSM in our discussion of topic 10, groundwater sustainability. We discuss the other three topics here.

Recommended Research: Residential Demand Management

Residential DSM policies fall into two broad categories: price and nonprice policies.

Price-Policy Research. According to economic theory, the lower the price of a product or service, the more of it people will demand. In Qatar, residential water is clearly underpriced: It is provided free to Qataris and is heavily subsidized for non-Qataris. Thus, Qataris have little incentive to conserve water. Revenues do not cover the costs of collecting, treating, and delivering water, let alone the costs of externalities associated with resource depletion or water and wastewater treatment. Thus, the first area of residential DSM on which Qatar should focus is in developing and testing the effectiveness of an efficient and equitable pricing system.

---

[69] Food and Agriculture Organization of the United Nations, 2008.

The most basic pricing policy is to charge consumers a uniform price for each unit of water used.[70] Extensive research on how residential consumers react to changes in the price of water has established that, under a simple pricing system, consumers do lower their water use in response to higher prices, albeit slightly. A meta-analysis of some residential water-demand studies indicates that the average price elasticity of demand for residential water is −0.41; in other words, a 10-percent increase in price is associated with a 4.1-percent decrease in water use.[71]

Water is a basic necessity for survival and, often even more than other basic needs, such as food or shelter, is perceived as having a "special significance."[72] Thus, proposals to increase water prices—for example, by shifting from fixed monthly charges to per-unit rates or increasing per-unit rates even slightly—are often fiercely resisted.[73]

One way in which some water utilities have tried to address the resistance to increasing water price is by adopting increasing-block pricing systems. In such systems, consumers pay a low price for the first few units of water, and higher prices for subsequent units consumed. The benefit is that all consumers can afford enough water to meet their basic needs, but high levels of use are discouraged. A challenge is that, in order to be effective, consumers must understand the prices they face, and several authors have argued that consumers do not fully understand complex pricing schemes.[74]

The resistance to water price increases is likely to be particularly true in Qatar, where the public has long received free or heavily subsidized water and might perceive such subsidies as a basic right. Using an increasing-block pricing system could allay some of those concerns by keeping the cost of using water for basic tasks, such as bathing and cooking, low. Thus, the institute should conduct research to determine

---

[70] Charging for each unit of water used, rather than simply charging a flat monthly rate regardless of amount used, requires that all households have water meters.

[71] Dalhuisen et al., 2003.

[72] Hanemann, 2006.

[73] Glennon, 2004.

[74] Liebman and Zeckhauser, 2004.

which pricing systems can be applied in Qatar, what the appropriate price of water should be in those systems, and how pricing can be made socially and politically acceptable.[75]

Nonprice-Policy Research. A second strategy for managing residential demand is to use nonprice policies. Rather than increasing the price of water regardless of what it is used for, nonprice policies target specific uses of water. A recent study suggests that routine activities, such as bathing, flushing toilets, and washing dishes, are responsible for most household water use in Qatar.[76] Therefore, targeting nonprice policies at specific activities that account for the largest shares of water use in Qatar, and identifying effective policies for lowering water use by those activities, could yield substantial water savings. For example, nonprice policies might include restrictions on watering lawns or washing cars on certain days of the week, or they might subsidize or require the installation of water-efficient appliances, such as low-flow toilets and showerheads.

Although nonprice policies are widely used, and existing research indicates that rationing and subsidizing water-efficient appliances can reduce demand, there is little evidence on the cost-effectiveness of specific nonprice policies. In Qatar, Kahramaa imposed a tax in 2007 on residential roadside water taps; since the imposition of a tax on these taps, many are still visible but not functional.[77] Qatari law also bans the use of potable water for washing cars or cleaning public yards by a hose or any other flushing tools. Those who violate this rule can face

---

[75] The economically efficient price of water—that is, the price that reflects the opportunity cost of water, or the value of the next-best alternative for which it could have been used—is its long-run marginal cost (LRMC). The LRMC is the cost of providing an additional unit of water, including the costs of collection, treatment, and distribution, and accounting for existing and future capital costs. Because the least expensive water supplies are exploited first, the LRMC of water generally increases over time. However, most water pricing does not cover the current cost of the delivered water, let alone its LRMC (Hanemann, 2006; Olmstead and Stavins, 2009).

[76] General Secretariat for Development Planning, 2009.

[77] "Preserving Water," 2010; these taps can be found on the outside walls of residential homes, for the public to use freely to fill water bottles or to drink. They are often seen as a charitable provision made by homeowners.

fines up to QR 1,000.[78] Similarly, it is against the law to leave pipes and other plumbing parts in disrepair, and Kahramaa advises the public to regularly check and repair plumbing and to report any leaks directly to Kahramaa.[79]

There is little evidence on the cost-effectiveness of specific non-price policies. In addition, DSM policies are often enacted during droughts and are bundled together. If price increases and numerous restrictions on water use occur at the same time, then it is difficult to determine which of the policies is actually responsible for a reduction in water use. Reactions to such policies are also highly context-specific and depend on household characteristics, such as income and household size, as well as climate patterns, such as rainfall and temperature.[80] Therefore, the institute should conduct research to understand both the costs and the potential benefits of specific nonprice policies given Qatar's climate, residential characteristics, and culture, all of which affect patterns of water use.

One potential avenue for conducting much-needed research on the effectiveness of price and nonprice policies would be to use experimental or quasi-experimental methods, which use elements of randomized controlled trials.[81] For example, rather than implementing a low-flow showerhead subsidy policy throughout Doha at the same time, the policy could be rolled out to different parts of the city in stages, thus allowing a comparison of water use in areas where the program was implemented earlier with use in areas where it was implemented later. This type of research would provide the institute with rigorous evidence on the effectiveness of specific policies in the Qatari context.

Recommended Research: Transmission and Distribution System
According to a report by Kahramaa, losses from the transmission and distribution system in Qatar account for up to 45 percent of water

---

[78] Law 26 of 2008 as reported in "Kahramaa Drive to Save Energy," 2010.

[79] "Kahramaa Drive to Save Energy," 2010.

[80] Dalhuisen et al., 2003; Mansur and Olmstead, 2007.

[81] See Nataraj and Hanemann, 2011, for an example of applying quasi-experimental techniques to water policy.

use.[82] If this is the case, then improving the performance of this system offers a significant opportunity for water savings. The institute should begin by working with Kahramaa to identify sections of the network that are responsible for water losses, as well as specific problems at each location. Certain issues, such as poor maintenance, might require the implementation of appropriate operation and maintenance scheduling, while others, such as the presence of corrosive soils surrounding the pipe, might require additional research on ways to minimize such corrosion. There is research on a wide range of issues associated with transmission and distribution systems, such as improving methods for minimizing corrosion or detecting leaks.[83] Once an assessment of the major sources of water losses in Qatar's network has been made, the institute should evaluate whether remedies that have already been developed elsewhere are appropriate or whether it would be worthwhile to conduct research on specific topics in this area.

Recommended Research: Water-Conservation Education

The institute should also extend existing programs to encourage everyone in Qatar to conserve water. Public education programs are often an important component of an overall DSM strategy that also includes price and nonprice policies. A public education program could disseminate information about water scarcity and specific water-saving tips through a variety of platforms, or it could offer information in a water report or bill about how a household's water use compares with that of the average use or with some target. Some programs already exist in Qatar. For example, Kahramaa currently provides, on its website, tools for calculating water use and tips for conserving water and undertakes advertising campaigns for conservation during the summer. QSTP, in conjunction with ConocoPhillips and General Electric, runs a water-conservation education center called the Global Water Sustainability Center, and Qatar has been involved in awareness campaigns during the GCC Water Week, an annual GCC-wide commemoration of

---

[82] General Secretariat for Development Planning, 2009.

[83] See, for example, EPA, 2010c.

World Water Day on March 22.[84] The institute should work with these institutions to undertake ongoing and active outreach and education campaigns.

Residential audits—in which water-conservation experts visit a home to assess water use, detect leaks, and provide low-flow devices—could also be offered.[85] Though it is difficult to separate the effects of public education from those of other water-conservation programs, one study indicated that public information campaigns were associated with water savings of 8 percent in various cities in California.[86] The main area of research for the institute would be to determine how to extend existing public education programs to maximize their effectiveness. The first step is to investigate water-use patterns: As discussed earlier, if there are certain activities (e.g., bathing) that account for most residential water use, then an education campaign will be most effective if it targets those activities—for example, by encouraging the use of low-flow showerheads. A second step is to work with public education experts to develop materials that are suited to Qatar's social and cultural needs and to determine the most-effective methods for disseminating information to the public.

Additionally, a conservation curriculum for primary and secondary schools could help to ensure that future generations understand the issue of water scarcity and embrace conservation and efficiency. Educational curricula are part of many public education campaigns.[87] Some institutions, including several universities, EPA, and water organizations, such as the Water Environment Federation, have developed curricula for incorporating water-conservation education into primary and secondary schools. There is some evidence documenting the impacts of education on water savings. For example, a study in central Jordan examined the effects that a water-conservation curriculum

---

[84] The Global Water Sustainability Center has laboratories for water research and hosts an interactive educational display on water conservation geared toward school children. Its website can be found in Global Water Sustainability Center, 2011.

[85] EPA, 2002.

[86] Renwick and Green, 2000.

[87] EPA, 2002.

had on high-school students who participated in eco-clubs. The study found that the curriculum increased students' knowledge about water conservation, as well as their reported practice of water-conservation measures.[88] The key challenge in this area is to adapt curricula designed in other countries for the specific cultural and social needs of Qatar's schools.

Collaboration Opportunities and Human-Capital Needs for Qatar

There might be ample opportunity for these researchers to collaborate with other institutions within Qatar. Collaboration with Kahramaa, in particular, will be necessary for addressing water losses, and such collaboration might also be helpful in extending Kahramaa's existing public education campaign. Additionally, BQDRI and Kahramaa recently announced plans to use the Qatar Sustainability Assessment System (QSAS), a green building rating system that emphasizes energy and water use.[89] Kahramaa and BQDRI will adopt QSAS standards to specify efficient plumbing fixtures, create a system for the collection and storage of rainwater, enable on-site treatment of water for future use, and create a landscape plan that reduces the need for irrigation.[90] Texas A&M University at Qatar also established the Qatar Sustainability Water and Energy Utilization Initiative in 2010, which provides scientific and technical research support and outreach activities with communities, industry, and the energy sector.[91]

In the GCC, SQU in Oman and King Faisal University in Saudi Arabia arguably have the most-robust water-management research programs, particularly for water management and conservation in agricultural production. UAEU is researching methods of using electronic

---

[88] Middlestadt et al., 2001.

[89] "Kahramaa, BARWA," 2010; the QSAS will use its rating system to measure the environmental performance of existing buildings in an effort to help ensure the maintenance of occupant health and environmental sustainability. There are more than 128 QSAS buildings registered in Qatar ("Kahramaa, BARWA," 2010).

[90] "Kahramaa, BARWA," 2010.

[91] The website for this initiative is Qatar Sustainable Water and Energy Utilization Initiative, undated.

sensors to monitor leaks and losses in the country's water system, and KAUST in Saudi Arabia is conducting water reuse research. A variety of institutions in the United States and Europe (e.g., the American Water Works Association and the National Resources Conservation Service of the U.S. Department of Agriculture) offer opportunities for collaboration beyond the GCC.

In order to carry out a research agenda in water DSM, the key resources needed are researchers in policy and economics, engineers (most likely civil and environmental) with experience in designing and operating water-delivery systems, and experts in developing and conducting public education campaigns. Qatar University offers undergraduate education in civil engineering, engineering management, and environmental sciences, which can build a foundation for local human-capital development in these areas. SQU offers Ph.D. programs in soil and water management and an M.Sc. in civil engineering with a focus on water resources. UAEU also offers an M.Sc. in water resources with follow-on Ph.D. opportunities in engineering. KAUST offers advanced degrees (M.Sc. and Ph.D.) in environmental science and engineering, which include a specialized environmental fluid mechanics and hydrology track. Finally, Masdar Institute of Science and Technology is offering an M.Sc. in water and environmental engineering, which can be the foundation for its interdisciplinary doctoral degree program.

Priority Topic 12: Integrated Water Resource Planning

Background and Motivation

Qatar currently faces multiple water-related challenges, including groundwater depletion, rising demand for urban water and sanitation, and increased water pollution. Qatar cannot address its water problems in a piecemeal fashion: Water production and consumption are intimately linked to other sectors, including households, energy, agriculture, and industry. Additionally, water-management decisions made today—e.g., about groundwater conservation—will likely have long-term consequences. Qatar must institutionalize ongoing IWRM that takes into account the relationships between demand from different sectors, different supply options, policies, and regulations, in both the near and long terms. In many ways, IWRM can be thought of as analogous to strategic energy planning, but for water.

IWRM is a holistic approach to water management that recognizes the links between different types of water sources, as well as the links between water management and other aspects of the economy. Although IWRM can mean different things to different users and in different contexts, a commonly used definition of IWRM, developed by the Global Water Partnership (GWP), is "a process which promotes the coordinated development and management of water, land and related resources, in order to maximize the resultant economic and social welfare in an equitable manner without compromising the sustainability of vital ecosystems."[92] Although the IWRM process will be unique to every individual country,[93] GWP has argued that a successful IWRM process involves three "pillars":

- Enabling environment: a set of policies, strategies, and legislation for sustainable water management
- Institutional framework: a framework for implementing the policies and strategies
- Management instruments: specific tools, such as DSM, that institutions can use to implement the policies.

We believe that IWRM could be valuable in Qatar for several reasons. First, there might be hydrologic ties between different sources of water, making their joint management necessary.[94] For example, in Qatar, the Doha groundwater basin is recharged not only from rainfall but also from infiltration due to landscape irrigation and leaking water pipes in Doha.[95]

Second, water policies could affect not only water resources but also other aspects of the economy. Land-use planning issues, in par-

---

[92] GWP, 2000.

[93] GWP, 2000.

[94] Ashton, Turton, and Roux, 2006.

[95] General Secretariat for Development Planning, 2009.

ticular, can be tightly linked to water management.[96] In Qatar, policies governing the use of groundwater are likely to affect many aspects of irrigated agriculture, such as cropping patterns and the extent of cultivated land. Moreover, different sectors compete for water, and supplying water in one sector could limit its availability in another. For example, Qatar's food security program seeks to substantially increase domestic food production. Despite advances in efficient irrigation or other technologies, agriculture is likely to be a major and potentially inefficient consumer of water, relative to its contributions to Qatar's economy. Nevertheless, domestic food production is a critical national interest. Research is needed to weigh these competing demands for water and coordinate long-term planning in water-intensive sectors.

Third, water management often falls under the jurisdiction of various institutions and involves multiple stakeholders, many of which might have different, sometimes conflicting, goals. IWRM seeks to engage all relevant stakeholders.[97] In Qatar, such stakeholders might include Kahramaa, the Ministry of Environment, farmers' groups, and industry representatives.

We recommend that the institute help Qatar assess and potentially adopt an IWRM process that is tuned to Qatar's needs. The institute should begin by conducting research to determine whether and how an IWRM would benefit water management and related issues in Qatar.[98] If the institute determines that IWRM would be valuable in Qatar, and the Qatari government agrees to embark on an IWRM process, then the institute should help identify an appropriate framework for implementing IWRM in Qatar. Because IWRM is a process, it is necessary to tailor the process to Qatar's resources, culture, society, and institutions.[99] Finally, if IWRM is adopted in Qatar, the insti-

---

[96] Mitchell, 2005.

[97] Ashton, Turton, and Roux, 2006.

[98] IWRM is not appropriate in every context. For example, IWRM might be appropriate for addressing complex water-management issues that involve multiple stakeholders, but it might not be appropriate for relatively straightforward, localized issues that fall under the jurisdiction of one institution (Hooper, McDonald, and Mitchell, 1999; Mitchell, 2005).

[99] Cardwell et al., 2006; Jønch-Clausen, 2004.

tute could undertake additional research to extend the basic IWRM framework.

Recommended Research: Determining Whether Integrated Water Resource Management Is Appropriate for Qatar

The institute should first research the value that Qatar would derive from adopting an IWRM approach to water planning. In particular, the institute could focus on identifying specific areas in which integration could be beneficial:

- Are there benefits to integrating the management of different sources of water (groundwater, desalinated water, reclaimed wastewater)?
- Which institutions have jurisdiction over water policies? Do their jurisdictions overlap, and, if so, are there any areas in which their missions lead to conflicting policies?
- Which institutions have jurisdiction over related policies (e.g., agriculture or land-use planning)? To what extent do water policies affect land use (and vice versa)?

If the institute finds that IWRM will benefit Qatar, its recommendations could help the Qatari government in deciding to adopt an IWRM process. In the next section, we discuss the major components of the IWRM process in more detail, with particular emphasis on tasks that the institute could undertake to facilitate the process.

Recommended Research: Tailoring Integrated Water Resource Management for Adoption in Qatar

The institute can next assist the government of Qatar in developing an IWRM plan, which lays out major elements of the IWRM process. In this section, we discuss potential key elements of an IWRM plan, adapted from the standards suggested by the California Department of Water Resources.[100] It is important to note that these elements comprise one possible set of components for an IWRM plan and that

---

[100]California Department of Water Resources (DWR), 2010.

Qatar's plan would need to be tailored to meet its specific institutional, societal, and hydrologic needs. The key elements of the plan are as follows:

- Governance. The IWRM plan should identify a group responsible for development and implementation of the IWRM process. The group might consist of several agencies, some of which have authority of water supply or water demand management (e.g., Kahramaa) and others that might be closely related (e.g., Ministry of Agriculture). The plan should also describe how major stakeholder groups will be involved, how decisions will be effectively made both inside and outside the group, how the group will communicate between members and with outside agencies and the public, and how the plan will be updated. The institute can assist the government of Qatar by identifying the relevant stakeholder groups that should be involved in the IWRM process and by identifying potential mechanisms that could be used to coordinate them.

- Regional description. The IWRM process should include a regional description in order to provide all stakeholders with information about current water-related conditions and to identify key issues to be addressed through the process. A description of Qatar should include key factors, such as the amount and quality of Qatar's water, habitats, protected land and marine areas, estimated water supply and demand from all major sectors over a 20-year horizon, and an assessment of current water quality. Qatar's social and cultural context should also be described, as should the relevant boundaries, both natural (groundwater basin boundaries, for example) and human-made (e.g., agency boundaries, urban areas). Major water-related issues and conflicts should also be discussed. Much of this information might become available through ongoing research conducted by the institute on water resource and environmental issues.

- Objectives. The IWRM plan should present clear objectives that address the water-related issues and conflicts and should prioritize the objectives. The objectives of the plan would be determined by the key stakeholders identified in the first step.

- Resource management strategies. The IWRM plan should document specific strategies that could be used to meet each objective. For example, water demand management might be addressed by encouraging low-flow irrigation techniques in agriculture or restricting outdoor water use in residential areas. Water supply might be increased by expanding desalination capacity or improving groundwater recharge. Water quality improvements might be achieved by improving water testing or treatment facilities. The institute can assist in developing specific strategies through its research on related areas, such as agricultural and residential water management and desalination.

- Impacts and benefits. The IWRM plan should indicate the potential impacts and benefits of each proposed project. The institute could provide input to this list based on its research on water and environmental issues.

- Project review process. The IWRM plan should document how specific projects will be submitted, reviewed, and selected as part of the IWRM process. The review process should identify how various members of the group responsible for overseeing the IWRM process would make decisions about project inclusion, and it should include a mechanism for stakeholders to provide input during the review and selection processes.

- Financing. The IWRM plan should identify sources of funding for all proposed projects.

- Performance and monitoring measures. It is important to develop a method of monitoring the IWRM process to ensure that progress is being made toward meeting the objectives of the IWRM plan and that the selected projects are being implemented.

- Stakeholder involvement. The IWRM plan should describe a process for allowing stakeholders to participate in the IWRM process on an ongoing basis.

- Coordination. The IWRM plan should identify opportunities and processes for coordination with existing water-management plans and land-use plans (both urban and agricultural), government agencies within Qatar, and other countries.

For these last five items, the institute could provide input by identifying processes, performance measures, and financing mechanisms that have been implemented in other countries and that could be adapted for use in Qatar.

It is not critical that Qatar's IWRM plan follow these specific guidelines but rather that it identify a process for coordinating among the relevant stakeholders to manage different types of water resources and to recognize the links between water policies and other sectors.

Recommended Research: Additional Research to Extend the Integrated Water Resource Management Framework

If Qatar decides to adopt an IWRM process, the institute could undertake additional research that extends the basic IWRM framework. In this section, we discuss two examples of such research.

One area of possible future research concerns uncertainty. Each stage of the IWRM process involves uncertainty that arises from several sources, such as incomplete or noisy data used to guide the process, an imperfect understanding of how natural and human systems work, discrepancy between different stakeholders over goals or outcome probabilities, and lack of knowledge of how future conditions will affect water management.[101] Recent research has developed new methods for dealing with certain types of uncertainty, including a robust decisionmaking method that identifies outcomes that perform adequately under a wide range of plausible future scenarios.[102] The institute could research ways to adapt Qatar's IWRM process to account for uncertainty due to climate change, uncertain water supplies, or other conditions.

Another future IWRM research area addresses water resource planning in urban settings. Most IWRM strategies tend to focus on river basins, and there is little understanding of how to implement

---

[101] Pahl-Wostl, 2007; van der Keur et al., 2008.

[102] Groves, Yates, and Tebaldi, 2008, p. 16.

IWRM strategies in urban contexts. Although considerable attention has been given to urban water management, most of this research focuses on specific water-management tools rather than on the development of an overall IWRM process.[103] The institute could conduct research on how best to integrate water-management planning within Doha—for example, on how to coordinate residential DSM policies with sanitary-system development.

Collaboration Opportunities and Human-Capital Needs for Qatar

To undertake research on IWRM, the institute will need researchers in a range of relevant disciplines, including public policy, economics, hydrogeology, and engineering (civil, environmental). Some of this expertise can be found in Qatar within the ministries, municipal governments, industry, and universities. In addition, many of the universities in the GCC countries offer advanced degrees (M.Sc. and Ph.D.) in water resource management, including SQU and UAEU.

There are opportunities for collaboration with other countries: Many other countries, including Egypt, Australia, Thailand, Vietnam, the western United States, and numerous European countries, have undertaken IWRM exercises. Coordination with other GCC countries might be particularly fruitful when it comes to addressing joint water concerns, such as shared aquifers. Qatar can also learn from recent IWRM activities in the UAE's Ministry of Environment and Water and Oman's Ministry of Regional Municipalities and Water Resources. In addition, organizations, such as GWP, have committed to working with countries to promote IWRM and have developed toolkits to assist in the IWRM process.

---

[103] J. Rees, 2006.

CHAPTER SIX

Priority Environment Research

Overview of Environmental Research

In the past several decades, Qatar has undergone sustained, high levels of population and economic growth. Although growth has significantly raised Qatar's national income, the combined forces of population growth, industrialization, and coastal development have adversely affected Qatar's environment. Sustainable development indicators from Qatar's Statistics Authority suggest that air-pollution levels in Doha are rising; many land areas are undergoing desertification; and environmental damage has been noted in Qatar's marine environment, including loss of biodiversity, coral-reef destruction, and an increase in coastal erosion and flooding.[1]

Qatar has taken several steps to address its environmental challenges, such as constructing wastewater-treatment plants, identifying several protected terrestrial and marine areas, and prohibiting shrimp fishing until stocks recover.[2] However, a great deal of uncertainty remains about the current health of Qatar's environment.

We recommend that the institute's environmental research focus, first and foremost, on characterizing the current state of Qatar's environment (topic 13). Once the state of the environment and the interrelationships between different ecological systems are understood, the institute can proceed to identify appropriate solutions for Qatar's envi-

---

[1] General Secretariat for Development Planning, 2009; Permanent Population Committee, 2010; Richer, 2008.

[2] General Secretariat for Development Planning, 2009.

ronmental challenges. The characterization exercise might indicate areas in which the institute could conduct research on innovative technologies or policy measures, but, in many cases, the solutions are likely to have been developed elsewhere, and the institute should focus on identifying the most-appropriate policies or technologies.

We also recommend that the institute undertake research that cuts across and informs other environmental, water, and energy issues (topic 14). This includes evaluating risks from climate change; using environmental economics to identify solutions to energy, water, and environment challenges; and undertaking a mass-energy balance to document the flows of energy and materials through the country or region.

Priority Topic 13: Environmental Characterization

The institute should characterize the state of Qatar's environment, including fisheries, the desert ecosystem, overall biodiversity and ecosystem services, groundwater resources, air and water quality, and waste-management practices. Following the initial characterization, the institute can then expand its research portfolio to identify and develop solutions for improving the state of the environment. In this section, we discuss several major aspects of the environment that should be included in the characterization. We have divided the recommended research topics into two broad areas: ecosystem and biodiversity characterization, and air and water quality and waste management. The first set of topics deals broadly with understanding ecosystems, while the second set addresses human-made pollution and waste. However, the two areas are linked: For example, coastal water pollution will affect the health of fisheries, so the research in these areas should be coordinated.

Recommended Research: Ecosystem and Biodiversity Characterization

Qatar's ecosystems contain at least 2,000 identified species, many of which have been threatened by recent development. In this section, we discuss recommendations for research to characterize fisheries and coastal ecosystems, as well as desert ecosystems, and to develop a broader understanding of Qatar's biodiversity and the services provided by its ecosystems.

Fisheries and Coastal Ecosystems. Fishing and pearling (pearl diving or hunting) have historically been important sources of livelihood in Qatar. The fishing industry in Qatar remains largely artisanal, and there are few suitable sites for commercial fishing. The number of fishing licenses is set at 515, but the size of fishing boats has grown over time, as has the total catch.[3] In an effort to protect its fisheries, Qatar has set up a Marine Resource Protection Committee that can monitor fishing activities and enforce licensing requirements.[4] Qatar has also banned shrimp fishing since 1993 in an attempt to allow the stock to recover; however, there has not been sufficient recovery, potentially because the shrimp stock is shared with Bahrain and Saudi Arabia.[5] The coastal ecosystem as a whole is also important: Nearly half the 2,000 known species in Qatar are marine species. As discussed in more detail in the section on water quality, coastal development and water pollution pose threats to marine creatures and their habitat.[6]

In order to promote the sustainable management of the fisheries that are a part of Qatar's heritage, as well as the health of other marine plants and animals, the institute should begin by documenting information about the coastal environment, including

- fish stocks and population trends
- the existing populations of other marine species
- the habitat needs of fish and other marine species
- total catch of each species in each season
- threats to marine species from coastal activities and pollution

---

[3] Food and Agriculture Organization of the United Nations, 2003; General Secretariat for Development Planning, 2009.

[4] "New Law to Help Boost Qatar's Marine Resources," 2010.

[5] Food and Agriculture Organization of the United Nations, 2003; General Secretariat for Development Planning, 2009.

[6] General Secretariat for Development Planning, 2009; Richer, 2008.

- the extent to which effects of dredging, reclamation, water intake, and pollution have already affected marine species and their habitats.

Using this information, the institute can conduct research to estimate the sustainable yield for each species—the total number of a particular species of fish that can be caught while maintaining the underlying stock. If needed, the institute can also research the types of policies that might be most effective in maintaining the catch of each species at or below the sustainable yield. This can include, for example, ongoing monitoring of certain marine species, seasonal restrictions on fishing, quotas on allowable catch per boat, and agreements with neighboring countries on total regional catch.

Desert Ecosystems. Desertification is defined by the United Nations Convention to Combat Desertification (UNCCD) as "the degradation of land in arid, semi-arid, and dry sub-humid areas . . . a gradual process of soil productivity loss and the thinning out of the vegetative cover because of human activities and climatic variations such as prolonged droughts and floods."[7] Increased groundwater use and intensive grazing reduce and change the types of plant cover, which makes land more susceptible to wind erosion, thus removing topsoil and decreasing fertility.

Qatar has grazing animals, including camel, sheep, goats, cattle, and horses; the density (the number of animals grazed on an area of land) of camels is the highest in the Middle East region. These grazing animals consume significant amounts of native vegetation, thus contributing to the disappearance of native species, such as the oryx and gazelle. In addition, the introduction of nonnative plant species that require irrigation increases groundwater use, which exacerbates the process of desertification.[8]

Qatar has already taken steps to protect its land by declaring five protected terrestrial areas.[9] To continue to address the challenge

---

[7] UNCCD, 2011.

[8] Richer, 2008.

[9] General Secretariat for Development Planning, 2009.

of desertification, the institute should begin by documenting baseline conditions, including

- the extent of desertification
- the populations and trends of grazing animal herds, as well as native species
- changes in vegetative cover over time, possibly using remote sensing techniques.

After this assessment, the institute can recommend potential goals for the health of Qatar's desert ecosystem. These goals could include the preservation (or reintroduction) of native species or the maintenance of plant cover. The institute can also conduct research on the most-effective policies or other methods for achieving these goals (e.g., limiting grazing areas, encouraging the use of different grazing-management methods). For example, a demonstration camel farm, sponsored by the United Nations Educational, Scientific and Cultural Organization, is currently being tested as an alternative to grazing camels on rangelands.[10]

Biodiversity and Ecosystem Services. Increased activity in and near Qatar's coastal and desert areas threatens marine and desert species. A closely related issue is the threat not only to individual species, such as shrimp or oryx, but to overall biodiversity and the ecosystem services provided by Qatar's environment.

Qatar's environment is host to a variety of terrestrial and marine organisms, many of which are uniquely adapted to survive under extreme conditions. Some species are considered endangered or threatened; however, data on known species are sparse, and new species are still being discovered.[11]

Some development activities pose threats to Qatar's terrestrial and marine species. Coastal development has the potential to destroy habitats, put stress on coral reefs, and damage fish nurseries. The pollution that industries discharge to the air and water can also harm species

---

[10] Richer, 2008.

[11] General Secretariat for Development Planning, 2009; Richer, 2008.

directly or through habitat damage. For example, mangroves, which stabilize coastal sediments and provide important habitat for many species, are affected by both coastal development and pollution. Moreover, tankers can carry invasive species, which compete with native species for dwindling habitat and resources, into the Gulf.[12]

In addition, continued development poses a threat to the ecosystem services that the environment provides. Ecosystem services are the benefits we receive from the natural environment. The Millennium Ecosystem Assessment (MEA) divides ecosystem services into four categories: provisioning (e.g., the provision of food, water, fuel), regulating (e.g., climate, flood, waste, disease regulation), supporting (e.g., nutrient cycling, soil formation), and cultural (e.g., spiritual, aesthetic).[13] Such ecosystem services can be affected by a variety of development activities. For example, intensive grazing leads to desertification, which reduces the fertility of the soil and, thereby, its capacity to provide food.

Qatar has taken some preliminary steps to assess and preserve its biodiversity and ecosystem services. As discussed earlier, it has set aside five terrestrial and three marine protected areas. The Qatar Bird Project, which is sponsored by the Ras Laffan Environmental Association, is assessing the bird population and identifying important bird habitat for preservation in Qatar. Qatar is also a party to conventions on species conservation, including the Convention on International Trade in Endangered Species of Wild Fauna and Flora and the Convention on Biological Diversity.[14] However, adequate data do not exist to assess the nature and extent of biodiversity loss.[15]

The institute should begin by conducting research to document current threats to Qatar's biodiversity and ecosystem services, including the following tasks:

- Identify all threatened or endangered species in Qatar, particularly those that are unique to the Gulf environment.

---

[12] General Secretariat for Development Planning, 2009; Richer, 2008.

[13] MEA, 2005.

[14] Richer, 2008.

[15] General Secretariat for Development Planning, 2009.

- Record population levels and trends in key marine and terrestrial species.
- Identify important habitat types and locations, and assess how ongoing and planned development is likely to affect important habitat.
- Document the sources and extents of threats to ecosystem services (e.g., the extent to which intensive grazing has affected the land's ability to provide food).

Following this assessment, the institute can then develop and recommend potential goals and policies for protecting Qatar's biodiversity and ecosystem services, which will depend on the specific nature of the threats identified. Both the assessment and policy research should be closely tied to research on the overall health of the coastal and desert areas.

Recommended Research: Water Quality, Air Quality, and Waste Management

As Qatar's economy and population has grown, its production of human-made pollutants and wastes has also increased. In this section, we provide recommendations for characterizing Qatar's current water quality, air quality, and waste-management practices. For air and water quality, the characterization might also include monitoring pollutant emissions and ambient pollution levels and developing models of pollutant fate and transport (the movement and chemical alteration of pollutants as they travel through the environment).

Coastal Water Quality. Qatar's coastline, including its islands, extends for more than 700 km.[16] The surrounding waters are shallow and highly saline and exhibit significant temperature variations.[17] The marine environment has not yet been mapped in detail.[18]

Coastal water quality is intimately linked with coastal ecosystems, which were addressed in the discussion of the subtopic on eco-

---

[16] General Secretariat for Development Planning, 2009.

[17] Richer, 2008.

[18] General Secretariat for Development Planning, 2009.

system and biodiversity characterization. In this section, we focus on assessing sources of pollution to coastal waters rather than on characterizing coastal ecosystems, but the research in these two areas should be coordinated.

There are two major sources of threats to the coastal environment: direct effects from land reclamation, dredging, and water intakes; and indirect effects from pollution. Some major construction projects, particularly along the eastern coast of Qatar, involve dredging (the removal of shallow sediments from underwater to, for example, create a deeper channel for shipping) or land reclamation (the creation of artificial land by dredging and infilling). Dredging and reclamation can reduce water quality, destroy habitat, create stress for coral reefs, and damage fish stocks.[19]

Electricity production, industrial processes, and desalination in particular have serious consequences for the health of the Gulf. Many power and industrial plants in the Gulf region use *once-through* cooling, in which significant amounts of water are withdrawn, passed through a cooling system once, and then returned to the Gulf.[20]

Withdrawing seawater for use in desalination and cooling causes impingement (when fish or other large organisms are trapped against intake screens) or entrainment (when small organisms are drawn into the intake pipes), leading to damage or death of marine plants and animals.[21] Additionally, because Gulf water is already at a high temperature, even more water must be withdrawn than in once-through cooling systems next to cold bodies of water. This means that the impact at intake is extensive. The water that is returned to the Gulf is heated and can contain metals from pipe systems. Both thermal and mineral pollution are contributing to the destruction of marine life in the Gulf.[22]

---

[19] General Secretariat for Development Planning, 2009.

[20] The Ras Laffan Industrial City common seawater-cooling system uses Gulf water and is intended to provide cooling capabilities for the growing number of industrial plants in the area. It includes once-through cooling systems and recirculation systems in which cooling water is reused (Pöyry Energy, undated).

[21] National Research Council, 2008.

[22] Lattemann and Hopner, 2008.

The health of coastal waters and marine life is also adversely affected by discharges from oil spills from production facilities and tankers, ballast water from tankers, and sewage from municipal treatment plants.[23]

Qatar has taken some steps to mitigate land-based pollution. It has constructed sewage treatment plants and adopted new technology to reduce the amount of chlorine in cooling water from colocated power and desalination plants. Three marine areas have also been identified for species protection. Nonetheless, some events indicating damage to the marine environment have been recorded recently: the destruction of corals reefs, an increase in red tides, and increased erosion and flooding, among others.[24] The institute can contribute to each of several necessary steps in protecting coastal waters. First, it can help establish a baseline for Qatar's coastal environment, including

- the extent of dredging and land-reclamation projects
- the types, amounts, and locations of pollutants entering marine areas
- the extent to which effects of dredging, reclamation, water intake, and pollution have already affected the coastal environment (e.g., erosion, flooding).

Second, the institute can help develop a monitoring system for pollution emissions, as well as ambient pollution levels, by taking the following actions:

- Identify key pollutants to be monitored.
- Assess the current ambient water–quality monitoring system for gaps in terms of pollutants monitored, areas covered, and monitoring schedule.
- Research pollutant-discharge monitoring and self-reporting policies for all major sources of water pollution (e.g., desalination plants, sewage-treatment plants).

---

[23] Khan, 2007; National Research Council, 2008; Richer, 2008.

[24] General Secretariat for Development Planning, 2009; Richer, 2008.

Once baseline pollution levels have been determined, the institute should third develop fate and transport models for key pollutants. The purpose of a fate and transport model is to estimate the movement and chemical alteration of pollutants as they travel through the environment. For example, chlorine that enters the sea in a cooling-water discharge can be dispersed over time, can volatilize into the air, or can undergo reactions with organic compounds. Researchers in many countries have developed fate and transport models for key pollutants in a variety of media, and those models can be adapted to suit the conditions present in Qatar's coastal waters.

The institute can recommend goals for the environmental health of Qatar's coastal waters. The goals can be used, in conjunction with the baseline analysis and fate and transport modeling, to recommend limits on pollution emissions and to identify appropriate technologies and policies for addressing coastal damage.

Finally, the institute should, in particular, seek ways of reducing the impact that electricity and water production has on the Gulf. This includes better siting and design of intake and discharge. It also includes making existing thermal power plants more water efficient without making them less efficient in producing electricity. Dry systems, for example, use air-cooled condensers to cool without water. Hybrid systems use air and water cooling in tandem. However, currently, these systems generally reduce the efficiency of the power plant, require additional space, and are noisy.[25] Second, water is a coproduct of oil and gas extraction, but this produced water contains chemicals and pollutants. This water must be treated and managed, but it could also be used in cooling. Research is needed to increase the tolerance that energy-production systems have to the impurities of such water.[26] Third, the institute should research ways to supplement existing generation supplies with low-water technologies, such as solar energy and fuel cell systems. Fuel cell and hydrogen systems consume an order of magnitude less water than traditional power plants.[27] Solar energy is

---

[25] U.S. Government Accountability Office, 2009.

[26] Electric Power Research Institute, 2007.

[27] Electric Power Research Institute, 2002b.

addressed in our discussion of topic 4, and fuel cells are addressed in our discussion of topic 5.

Groundwater. Groundwater is Qatar's main source of freshwater and is an important element of water security in Qatar. Groundwater, which is almost entirely used for agriculture, is extracted four times more quickly than it can be recharged by rainfall.[28] The quality of groundwater is also declining over time. As groundwater is depleted, seawater intrudes into the shallower aquifers, as does more-saline water from the underlying aquifers. Irrigation water that percolates back down to the aquifers also dissolves salts as it passes through the soil, further increasing salinity.[29] Moreover, the use of fertilizers, which are likely to contribute to chemical contamination in groundwater, is rising.[30] Thus, groundwater-sustainability research is critical to maintaining this resource and attaining water security in the country. Given its importance, we address research related to groundwater sustainability in our discussion of topic 10.

Air Quality. Outdoor air-quality monitoring is conducted at several stations in Doha and Al Khor. Available data from the Qatar Statistics Authority indicate that concentrations of particulate matter of less than 10 microns in diameter (PM-10) exceeded allowable limits in 2008.[31] Outdoor air pollution has a variety of human health and environmental consequences, including increased incidence of respiratory problems, reduction in visibility, and crop damage. PM-10 in particular is linked to respiratory problems.[32]

Indoor air quality can also have significant effects on human health. Some of the most-important sources of indoor air pollution are wood- and coal-burning stoves, household cleaning products, and

---

[28] Food and Agriculture Organization of the United Nations, 2009.

[29] Food and Agriculture Organization of the United Nations, 2009.

[30] Permanent Population Committee, 2010.

[31] General Secretariat for Development Planning, 2009; Permanent Population Committee, 2010.

[32] EPA, 2010a.

insulation that contains asbestos. Outdoor air pollutants, such as radon and pesticides, can also infiltrate homes.[33]

In order to address air-quality issues, the institute should conduct research that assesses baseline levels of key pollutants, including the following tasks:

- Determine key outdoor air pollutants to be monitored.
- Research the current air-quality monitoring system for gaps in terms of pollutants monitored, areas covered, and monitoring schedule.
- Identify key sources of indoor air pollution.

As with coastal water quality, the institute should develop fate and transport models for important outdoor air pollutants, recommend potential ambient air–quality goals, and conduct research on monitoring and self-reporting policies for major sources of air pollution. Some progress has already been made on this front: Several air monitoring stations are operating in Doha and Al Khor, and Qatar Petroleum, along with outside experts, has developed a regional model of surface ozone for Qatar.[34]

EPA recommends three strategies for addressing indoor air pollutants:

- Eliminate the source of pollution.
- Improve ventilation.
- Use an air cleaner.[35]

If the baseline assessment identifies indoor air pollution as a key issue, the institute should conduct research to determine the most-effective ways of dealing with the specific problems identified. Since the quality of indoor air and the best strategies for improvement are

---

[33] EPA, 2010d.

[34] General Secretariat for Development Planning, 2009; Al-Mulla, Ahmed, and Lecoeur, 2009.

[35] EPA, 2010b.

likely to differ across households, an education campaign to teach households how they can improve their indoor air quality might be most effective.

Waste Management. Waste management encompasses the collection, treatment, and disposal or recycling of solid waste from domestic, commercial, and industrial sources. Waste management presents some environmental challenges. The first is ensuring that waste is disposed of in landfills or other appropriate facilities rather than accumulating or being deposited in areas where it could cause a public health or environmental hazard. A second challenge is efficiently dealing with different types of waste—for example, safely handling and disposing of hazardous waste and recycling certain materials. Third, once the waste is contained in a landfill, it must be carefully managed by means of a bottom liner, a cap, and possibly leachate- and gas-collection systems, to ensure that, as the waste degrades, hazardous liquids and gases do not contaminate the surrounding soils, groundwater, or air. Fourth, an important component of waste management is developing a program that allows certain wastes to be recycled when doing so can yield economic or environmental benefits.

Qatar has already taken actions to improve waste management. For example, an integrated waste-management facility is being built south of Doha for both residential and commercial solid waste. The plant should allow for separation of organic matter for composting, as well as high-energy waste for incineration, in a waste-to-energy plant.[36]

To address the environmental challenges associated with waste management, the institute should conduct research to determine whether the current waste-management system addresses the following issues:

- providing adequate waste-collection systems for all households, commercial buildings, and industries
- appropriately handling and disposing of hazardous waste from residential, industrial, and commercial users

---

[36] "Qatar," 2008.

- constructing and operating waste-management facilities to adequately manage leachate and gas emissions from the landfill and to avoid public health risks from exposed waste
- creating additional infrastructure for waste collection and management to meet the needs of the growing population
- developing processes and facilities for recycling materials when recycling is more economically or environmentally desirable than disposal, and building awareness of such recycling programs.

The institute can then recommend solutions to address specific gaps identified through this assessment. As discussed in more detail in the next section, some policy and technology solutions for waste management already exist, and the most-appropriate research for the institute might be to identify existing solutions that should be adapted for use in Qatar.

Next Steps: Identifying Policy and Technology Solutions

For each environmental challenge that is identified through the characterization process, the institute should conduct research to determine the most effective way of addressing that challenge, whether through technology or policy solutions.

Some solutions will require the selection of the most-appropriate technology. Some mature technologies are available for reducing air-pollution emissions, for mitigating the effects that water intake and concentrate discharges from desalination plants have on marine life, and for managing wastes. Qatar has already adopted the best available technology standard of pulse chlorination to reduce the chlorine discharged into the sea with cooling water.[37] Similarly, once the institute has helped Qatar develop goals for maintaining or improving the state of its environment, the institute should conduct research to identify which specific technologies would be most effective in meeting those goals.

The institute should also evaluate Qatar's current strategies for procuring technological solutions to environmental problems. In par-

---

[37] Richer, 2008.

ticular, the institute should help Qatar develop a process for identifying the most-appropriate pollution-prevention technology, as well as the skills and training that workers will need in order to install and operate it. Research on different types of contract vehicles could also be helpful in ensuring that Qatar obtains the technology in a cost-effective manner.

If a policy solution is considered most appropriate for meeting a particular goal, the institute should conduct research on how proven policies from other countries could be adapted to suit Qatar's needs. For example, monitoring and self-reporting requirements for stationary air and water emission sources are used in many countries and can be transferred to Qatar. Other policies, such as a requirement for habitat restoration, or the development of a market for pollution trading, will require adapting strategies used in other locations to specific conditions in Qatar.

Collaboration Opportunities and Human-Capital Needs for Qatar

The institute's environmental characterization research should involve stakeholders, including the Ministry of Environment, the Qatar Supreme Council of Health, and the QNFSP. There might also be opportunities for the institute to collaborate with other research institutions. For example, ExxonMobil Research Qatar recently announced a program to identify native plants that can clean industrial water.

There are also many opportunities for collaboration with neighboring states. The health of Qatar's coastal waters, fishing stock, and air are affected not only by actions taken within Qatar but also by actions of neighbors, including Saudi Arabia, Bahrain, and the UAE. Thus, coordinating efforts with neighboring countries can increase the effectiveness of pollution-monitoring networks, fate and transport modeling, and fish stock management.

Kuwait University and KISR are doing a large amount of work on air-quality assessment and monitoring of marine pollutants and their effects in the coastal regions, including a study by Kuwait University of the impact of Doha's west desalination plant cooling water in Sulaibikhat Bay, Kuwait. KISR has also developed marine baseline surveys and the most comprehensive radiological survey in the region. SQU in Oman is working on a database for the Arabian Gulf ecosystem and has cataloged a baseline biodiversity index for Oman for conservation purposes. UAEU has ongoing research in environmental impact assessments and environmental health strategies and is in the process of developing an electronic portal for environmental data across the UAE and a geographic information system database for water resources.

To undertake ecosystem and biodiversity characterization, the institute will need to draw on the expertise of biologists and ecologists, while the water-quality, air-quality, and waste-management research will also require environmental scientists and engineers. These capabilities can be found or developed regionally. Kuwait University has a depth of environmental research expertise and offers opportunities to develop environmental scientists through degree programs in its Department of Earth and Environmental Sciences.[38] SQU, which also conducts extensive environmental research, offers focused Ph.D. programs in marine sciences and fisheries, as well as soil and water management, and an M.Sc. program in natural-resource economics.

Priority Topic 14: Crosscutting Environment Research

In addition to undertaking research in the specific areas laid out in the discussion of topic 13, the institute should perform research in three crosscutting topics that are applicable to Qatar's environmental challenges and are linked with its economic growth. The first topic— creating a mass–energy balance of Qatar's economy—involves developing an understanding of how different industrial activities interact with each other and with the environment. The second topic—considering the potential impacts of climate change—is relevant to all aspects of the environment and to Qatar's economy. The third topic—environmental economics—addresses ways in which economic tools can be used to conduct cost-benefit analyses for potential environmental solutions and to minimize the costs of reducing pollution.

---

[38] This program offers degrees in geology, desert studies, and marine and environmental sciences.

Recommended Research: Mass–Energy Balance and Industrial Ecology

As addressed in our discussion of topic 13, a key component of the environmental characterization is for the institute to document pollutant emissions to air and water. In addition, the institute should use data on pollutant emissions, along with information about energy and material flows, to provide information that researchers can use to develop a mass–energy balance for Qatar.

Developing a mass–energy balance is part of a growing field known as *industrial ecology*. Although the concept of industrial ecology is still emerging and there are many definitions, the *Journal of Industrial Ecology* article defines it as a

> field of science that systematically examines local, regional, and global flows of materials and energy in products, processes, industrial sectors, and economies. It focuses on the role of industry in reducing environmental burdens throughout the product life cycle from the extraction of raw materials, to the production of goods, to the use of those goods, and to the management of the resulting wastes.[39]

The basic concepts behind industrial ecology include the idea that the interactions between different industrial activities and between industrial activities and the environment should be recognized, that waste from one process should be used as an input for other processes, that environmental impacts from industrial processes should be minimized, and that industrial processes should be integrated with, and should emulate, natural processes.[40]

The institute should begin by developing a mass–energy balance for Qatar. A mass–energy balance documents the flows of energy and materials throughout a system. If the system is considered to be the entire country of Qatar, then inputs would consist of domestically extracted, as well as imported, energy and materials. Outputs from the system would consist of exports, as well as wastes and losses

[39] Lisfet, 1997.

[40] Keoleian and Garner, 1995.

from the system. The institute can develop input-output tables that document flows between sectors within the system. An input-output table for an economy shows how the outputs of one industry are used as inputs by other industries. Traditionally, input-output tables were used to map the monetary values of the flows of goods and services; however, they have more recently been used to estimate energy flows between industries as well.[41]

The mass–energy balance can then be used for several purposes. First, it can be used to characterize the types of resources used by each sector. Second, it can identify specific industries and products that use large amounts of energy or water, so that research efforts can be focused on developing ways to reduce energy or water use in these areas or to shift production or demand toward less energy- or water-intensive activities.[42] Third, the mass–energy balance can identify activities that are associated with high levels of pollution, thus suggesting potential industries in which to concentrate research on pollution control. Finally, the mass–energy balance can identify by-products from certain processes that might be used as inputs to another process, thereby improving efficiency and reducing the overall waste generated by the system.

Studies that utilize mass–energy balance estimates are typically undertaken by academic or government researchers with the intent of improving efficiency of, or reducing pollution damages from, a process or economic sector in a country.[43] They are routinely carried out in OECD countries and have their own professional societies, journals, and International Organization for Standardization (ISO) standards. For Qatar, the first step would be to improve data quality and access to material and energy flows in the country, so that these analyses could

---

[41] Bin and Dowlatabadi, 2005; Park and Heo, 2007. For example, CMU has developed an Economic Input-Output Life Cycle Assessment for the United States, which provides estimates of the economic value, energy and water requirements, and pollutant emissions for nearly 500 activities (CMU, undated).

[42] The energy used to make a product is sometimes referred to as *embodied energy*, while the water used to make a product can be referred to as *virtual* or *embodied water*.

[43] Some of these studies fall under the category of life-cycle assessments.

take place. In the United States, the data that are utilized for such studies come from the EPA toxic-releases inventory, the Commerce Department's economic census and input-output accounts, the Department of Energy's Energy Information Administration, the Department of the Interior's mineral analyses, and various other sources. We recommend that the institute first measure Qatar's use of materials and energy, then make the data available, so that researchers, both internal and external to the institute, could undertake such analyses.

Recommended Research: Climate Change Impacts on the Environment

Evidence from the Intergovernmental Panel on Climate Change indicates that the average global temperature could increase between 1.1 and 6.4 degrees Celsius within the next century.[44] Many of the facets of Qatar's environment discussed earlier—coastal waters, fisheries, desert lands, air quality, and biodiversity—can be affected by climate change.

Even if carbon emissions are reduced in the coming decades, it is likely that current temperature trends will continue until at least 2030. Climate change is predicted to reduce food and water security, raise sea levels, damage ecosystems, and harm human health. Moreover, the effects of climate change are likely to differ by region; for example, climate change might increase the potential for agricultural output in industrialized countries but decrease it in the MENA region.[45] Therefore, it is important that the institute conduct research on the potential effects of climate change on each of the facets of its local environment discussed earlier.

Recommended Research: Environmental Economics

Many of the regulations and technologies that Qatar selects to address its environmental challenges are likely to be costly. For example, there are mature, proven technologies to reduce the amount of pollutants discharged to the air and water, but many of them require the addition of equipment to existing industrial processes. Cost–benefit analy-

---

[44] United Nations Development Programme (UNDP), 2007.

[45] UNDP, 2007.

sis could be helpful in determining which solutions to adopt. Since many of the benefits of environmental technologies or policies involve improvements to human health or the environment, it is often necessary to translate these benefits into monetary values so that they can be included in such an analysis.

The valuation of environmental goods and services is challenging because it involves estimating values that are not observed in market transactions. For example, the direct value of a product is generally its market price. However, environmental goods and services usually do not have a market price. Rather, their value is often an indirect value (e.g., ecosystem services, such as nutrient cycling or climate regulation), a nonuse value (e.g., the value people place on the existence of forests, even if they never visit them), an intrinsic value (e.g., the value of a particular species regardless of whether humans value its existence), or a value to future generations, known as a *bequest value*. Environmental economists have developed a set of tools for estimating the values of environmental goods and services, as well as valuing the damage caused by environmental degradation.[46] The institute should use these existing tools to conduct research on the valuation of potential benefits of specific environmental technologies or policies that are being considered for Qatar.

The institute should also conduct research on the potential use of market-based tools and policies to achieve an environmental goal in the most efficient way. For example, if two factories release the same type of pollution into the air, and the goal is to cut total emissions by a certain amount, a traditional, "command-and-control" regulation would require that both factories reduce their pollution to the same level. However, an alternative solution would be to place a tax on pollution or to issue pollution permits to both firms and allow them to trade between themselves (*cap and trade*). In the cap-and-trade case, if it is relatively inexpensive for the first firm to reduce pollution, it can reduce emissions to below its permitted level and sell the extra permits to the second firm. Both taxes and cap-and-trade mechanisms reduce the overall cost of pollution control by encouraging firms that will

---

[46] Smith, 1996.

incur the least cost to reduce pollution to the greatest extent. Well-known pollution-trading schemes include the market for sulfur dioxide in the United States and the market for GHGs in the European Union.[47]

Designing a market for trading pollution credits requires a substantial amount of knowledge about local baseline conditions, as well as an understanding of the fate and transport of pollutants. For example, in the scenario in the preceding paragraph, it is important to understand whether emissions from the two plants have the same effect on ambient air quality or whether emissions from one plant would be more harmful to human health or the environment (which might be the case if, for example, one plant were located close to a residential area). Therefore, developing a baseline characterization of the environment is a critical prerequisite for considering market-based solutions to environmental challenges.

Collaboration Opportunities and Human-Capital Needs for Qatar

Opportunities for collaboration might exist within Qatar and in the region. For example, the institute should work closely with Qatar's Ministry of Energy and Industry and Ministry of Environment to collect data that are needed to develop a mass–energy balance. Abroad, researchers at King Faisal University in KSA are researching the effects that climate change has on agricultural water requirements.

Research in these areas will need to draw on a variety of disciplines, including climate scientists, environmental scientists and engineers, and researchers in environmental economics and policy. As noted in the discussion of environmental characterization (topic 13), these capabilities can be found or developed regionally—for example, at Kuwait University, which offers degree programs in its Department of Earth and Environmental Sciences, and SQU, which has an M.Sc. program in natural-resource economics.

---

[47] European Commission Directorate-General for Climate Action, 2010; EPA, 2005.

CHAPTER SEVEN

Additional Insights on Other Research Topics

In the process of identifying priority research topics for the institute, several insights emerged about other research that might be beneficial for Qatar. Such research includes secondary topics that offer opportunities and benefits for Qatar but that do not strongly address Qatar's most-pressing sustainable development needs:

- marine-algae biofuels
- advanced hydrogen
- polymers, aluminums, and plastics
- closed agriculture systems and advanced aquaculture.

This other beneficial research also includes crosscutting and collaborative research topics that are potential opportunities for future collaboration or are of common concern to the Environment and Energy Institute and the other research institutes that QF is considering:

- environmental health
- occupational health
- food safety
- ultrapure water
- materials
- social science.

Because these insights are by-products of the main analysis, they should be thought of as illustrative rather than exhaustive and are not presented in any particular order.

Secondary Research Topics

Marine-Algae Biofuels

Algae are abundant all over the world and can grow in fresh, ocean, and waste water. About half of their mass is oil, which can be converted into ethanol or diesel and used as fuel in existing systems, such as airplanes and automobiles. As a result, algae biofuels have received significant attention as a potential substitute for fossil fuels.[1]

Marine-algae biofuels might be a high-risk, high-reward future area of research for Qatar. Ocean algae would use two of Qatar's most-abundant resources—solar energy and its Gulf water—and create a product for which there is already a large market.

Hydrogen Economy

Hydrogen fuel cells use hydrogen as the energy carrier to produce electricity. This offers numerous advantages over the combustion of fossil fuels: Hydrogen fuel cells do not emit conventional pollutants, they do not produce GHGs, and they allow distributed production. Thus, a hydrogen economy—one in which such fuel cells are the primary source of energy (e.g., replacing combustion engines in vehicles)—would have environmental sustainability benefits.

Hydrogen production, transportation, storage, and distribution are key hurdles to a hydrogen economy, but Qatar has advantages in overcoming these hurdles. Qatar is well positioned to produce hydrogen given its abundant natural gas resources. Hydrogen is also produced in the petroleum-refining process. Qatar is also a small country and could deploy hydrogen infrastructure more easily than larger countries could. Thus, hydrogen research, such as improving the pro-

---

[1]   Gross, 2008.

duction and transportation of hydrogen, could be promising for Qatar and could become a future subtopic for a fuel cell research program.

Polymers, Aluminums, and Plastics

The production of polymers, aluminums, and plastics requires fossil fuel feedstocks and significant amounts of energy. Given that these are abundant in Qatar, new industries could develop in these areas. Research would help to develop more-efficient methods of producing polymers, aluminums, and plastics and lay the foundation for such industries.

Closed Agriculture Systems and Aquaculture

Qatar has a small agricultural sector and imports much of its food. Thus, food security is a key national concern, and Qatar has created the QNFSP with the objective of increasing domestic agricultural production and strengthening the security of food imports to diminish food supply deficit. Despite being small, Qatar's agriculture sector has a significant impact on Qatar's groundwater (see topic 10). Thus, part of the QNFSP's mission is to use desalinated water instead of groundwater in order to prevent further degradation of groundwater. It also seeks to optimize irrigation and agricultural operations to improve water efficiency.[2]

Closed agriculture systems are self-contained and recycle air, water, and other resources throughout the system. This reduces the overall amount of resources needed. Research into closed-system agriculture designs tailored to Qatar's resources might offer ways of further reducing water and resource needs in Qatar's agriculture sector. Sustainable aquaculture—the farming of marine plants and animals under controlled settings with emphasis on sustainability—might offer another opportunity to address Qatar's food security issues while also addressing marine ecosystems. Research into aquaculture for the Gulf might therefore be valuable for Qatar.

---

[2] The QNFSP's mission statement can be found at QNFSP, undated.

Crosscutting and Collaborative Research Topics

Environmental Health

Environmental health addresses the environment's effects on the health of the general population. For example, respiratory illness has been linked to poor outdoor air quality; "building sickness" has been linked to poor indoor air quality.[3] The Environment and Energy Institute and a health research institute could collaborate on research to assess such environmental health impacts and to assess and recommend mitigation options.

Occupational Health

Occupational health is related to environmental health: It is concerned with the safety, health, and well-being of workers and employees in particular. This includes the impact that air quality, water quality, and the environment have on workers. This research is multidisciplinary and presents another opportunity for the Environment and Energy Institute to collaborate with a health research institute, particularly on such issues as the role that green buildings play in worker health.

Food Safety

Food-safety research is concerned with illnesses associated with the food production cycle, from bacterial contaminants to pesticides to soil pollutants. This is another potential area for collaboration between a health research institute and the Environment and Energy Institute because it involves research on the fate and transport of pollutants that can contaminate the air, land, and water.

Ultrapure Water

If Qatar's computing facility includes fabrication research, it might have a need for ultrapure water. If the environment institute is simultaneously undertaking research to make desalination more efficient, the two institutions could collaboration on ways of making desalinated ultrapure water more efficient.

---

[3]   Pervin, Gerdtham, and Lyttkens, 2008; EPA, 2010b.

Materials Research

Many of the priority research topics we recommend involve materials research, e.g., solar energy, desalination, fuel cells. It is likely that health, computing, and other technical research areas will also draw on materials research. Thus, materials research might be a crosscutting research program across many of the technical research institutions that QF is considering.

Social-Science Research

Similarly, many of the recommended priority research topics involve social-science research. For instance, water demand management requires research on the residents' responses to different policy options. Green building research must consider how people live and work in buildings and how new designs could affect cultural and social norms. Indeed, each of QF's institutions seeks to conduct research in a socio-technical area, so a social-science research track that addresses social and decisionmaking issues in and across each area would provide an important basis for collaboration.

CHAPTER EIGHT

Recommended Next Steps for the Qatar Foundation

Qatar's leadership has created a vision of sustainable development for the country, and the establishment of a national research institute is an important step in realizing this vision. We identified 14 priority research topics in the areas of energy, water, and environment that the Qatar Foundation should consider for this institute. These are listed in Table 8.1.

These priorities are only the beginning of a journey to realize such a bold vision. We recommend that QF develop an initial research portfolio for the institute from the list of priority topics identified in this research.

Select Initial Research Topics

The institute's portfolio should consist of only two or three topic areas in the initial two to three years. This controlled growth allows some diversity in research capabilities from the beginning while also enabling the institute to mature and develop a governance approach, funding mechanisms, an institutional culture, partnerships, and other facets. The two to three initial research topics will serve as a test bed for this institutional framework.

A few criteria should be considered when selecting these initial research topics. QF should choose topics

- of greatest importance to Qatar's energy and environment that could be applied in the near term, so that the institution under-

Table 8.1
Priority Research Topics for the Institute

Area	Topic
Energy	1. Natural gas production and processing
	2. Petroleum production and processing
	3. CCS
	4. Solar energy development
	5. Fuel cells
	6. Green buildings
	7. Smart grids
	8. Strategic energy planning
Water	9. Desalination
	10. Groundwater sustainability
	11. Water demand management
	12. IWRM
Environment	13. Environmental characterization
	14. Crosscutting environmental research

NOTE: Numbers correspond to topic numbers in this book and are not intended to convey any sense of priority among the topics.

takes research that meets Qatar National Vision 2030 goals from the beginning

- in which Qatar already has or can quickly recruit researchers with the necessary expertise
- in which Qatar already has or can quickly develop the laboratories or other facilities for undertaking the research
- that are not redundant with research being done in the region or elsewhere or in which Qatar has a strategic advantage.

Our research principally offers guidance on the importance of a topic to Qatar's energy and environment goals, and it offers insights about human capital and about research being conducted in the region. In our research, several topics stand out as strong candidates for these initial topics. These are natural gas research; desalination, ground-water, and water demand–management research; and environmental characterization.[1]

Natural gas (topic 1) is key to Qatar's continued economic development. It is a longer-term resource than petroleum and has many technological and environmental advantages over petroleum and other fossil fuels. Qatar also ranks third globally in natural gas reserves, which suggests that Qatar should be a leader in research in natural gas research.[2] For these reasons, natural gas research is of utmost importance to Qatar. Qatar is already undertaking natural gas research, which suggests that the institute can recruit the needed talent and could have access to the necessary facilities, perhaps by collaborating with other institutions in Qatar. Simultaneously, the institute must be careful not to duplicate natural gas research already being conducted in the public and private sectors in Qatar and should consult with other research institutions to identify specific subtopics that are complementary. Other countries in the region and abroad are undertaking natural gas research, but Qatar's strong natural gas industry and resources suggest that Qatar has a strategic interest and advantage in being at the forefront of this work.

Water is a critical resource in Qatar, and Qatar's growth and security depend on having reliable sources of clean water for its population and industries. Research in desalination (topic 9), groundwater sustainability (topic 10), and water demand management (topic 11) seeks to balance the supply and demand of water. We recommend that the institute undertake research in at least one of these three areas, perhaps as part of an initial, integrated water-research program. Water security is a global concern, and many countries in the region and around

---

[1] Although we have identified several potential partners for collaboration during this study, efforts to establish partnerships should be made deliberately and largely after primary decisions about research priorities have been made. In this manner, collaboration can be made specific, with consensus about research goals and resources.

[2] Energy Information Administration, 2011a.

the world are undertaking water research. Nevertheless, groundwater sustainability and water demand management are important national issues; research undertaken in Qatar for Qatar is needed to address them. Much desalination is being done elsewhere, but Qatar's rapid growth and interests in food security suggest that it might be in Qatar's national interest to accelerate research in this field. Given Qatar's high levels of sunlight and long history of operating thermal desalination plants, it could take a leadership role in optimizing and scaling up solar-powered desalination plants in particular. Our research suggests that Qatar has some human capital in these areas, but more talent will need to be recruited, possibly from the regional institutions offering advanced degrees in these fields.

Qatar's environment is under great stress. The loss of biodiversity; pollution in the air, water, and land; and high resource use fundamentally threaten the sustainability of Qatar's development and its ability to achieve *Vision 2030* goals. Additionally, Qatar's environment will continue to deteriorate, possibly beyond repair, unless counteractions are taken. Thus, environmental characterization (topic 13) in the near term is of utmost importance to Qatar. As with water research, environmental research is being done globally, but research in Qatar and for Qatar is needed to address many of Qatar's own environmental challenges, which require policy and other interventions. Our study did not indicate that environmental research is currently a key area of research in the country. However, environmental research could present an important opportunity for collaboration with neighboring countries that share Qatar's concerns and that have knowledge of shared resources, such as Gulf water and ecosystems. Some regional universities offer advanced degrees in such fields as marine sciences and soil and water management, from which Qatar can draw talent.

These topics are initial recommendations based on the findings of this research. However, QF should undertake its own planning exercises to assess human-capital availability in detail, select subtopics, and determine the necessary investments in facilities. This process could be informed by cost–benefit analyses or other quantitative methods to weigh investments against perceived benefits.

Create Conditions for Continued Growth

QF should also create conditions for this research portfolio to grow and address many of these priority research areas. This can be done in the near term by developing a five- to ten-year road map for expanding the institute's research capabilities into other areas. This can begin during the process of selecting the initial research topics. Those topics that are ranked strongly but not chosen as the initial two to three topics are likely candidates for near-term growth. The mapping process should be ongoing during the first few years of the institute's development, considering the criteria given in the previous section, synergies between proposed and existing topics, and lessons learned.

This top-down mapping can be complemented by a bottom-up approach to growth in which investigators suggest research topics, submit proposals, and compete for supplementary funding from the institute. Such an approach draws on the talents and knowledge of existing researchers, creates opportunities for more-junior staff to lead research, can serve as a form of peer review and quality assurance, and grows the program organically. Because it requires a modest base of researchers, this mechanism might be feasible only after the first few years.

The top-down and bottom-up growth mechanisms can be very complementary. The top-down road map can identify primary priorities, while the bottom-up supplementary funding can spark research in secondary areas. Alternatively, the top-down can identify broad research areas (e.g., solar energy), and the bottom-up can identify specific subtopics (e.g., dust-resistant glass or hybrid solar/natural gas power plants).

Track Progress

This endeavor is ambitious. It will require a long-term commitment by QF. As we do for the research it will be sponsoring with the institute, we recommend that QF should approach this overall endeavor as an experiment in itself, in which decisions to invest in a research area are treated as hypotheses that can be tested and altered as new evidence is available. The hypothesis is that the recommended research topic will yield benefits to Qatar at an investment cost that is reasonable.

This hypothesis can be tested by estimating potential outcomes and comparing them with evidence that emerges as research is funded and undertaken. We recommend that the institute focus on at least two kinds of outcomes that reflect the institute's mission: (1) the extent to which the institute is furthering Qatar's energy and environment goals and (2) the extent to which it furthers Qatar's goals of becoming an international leader in research.

This evidence, as with any experiment, can be measured continuously by asking several questions, such as these:

- Is the investment supporting research that is both cutting-edge and unique to Qatar?
- Is the research producing results that are well respected by peer reviewers?
- Is the research producing results that further Qatar's vision of sustainable development?

Metrics and thresholds to answer these questions should be developed as part of the mapping and strategic-planning processes, but the slate of research topics presented here provides seeds for this endeavor to grow.

Potential Research Topics and Subtopics

Table A.1 catalogs potential research topics that emerged from RAND expert panel discussions in step 1, generating a collection of topics, in the methodology for identifying priority research topics and the subtopics in each area. This list was used to seed discussion during the panels' assessment activities in step 4, which was to select priority research topics in panels (see Figure 2.1 in Chapter Two).

Table A.1
Potential Research Topics

Topic Area	Topic	Subtopic
Energy sources	Solar	Solar thermal systems, hybrid systems (PV/wind, solar thermal/natural gas), PV, passive solar, solar desalination
	Wind	Turbines, aeronautic blades, grid integration, system integration
	Natural gas	Reservoir characterization and engineering, natural gas drilling and recovery, cost and emission control for natural gas processing (including LNG and GTL), remote or microprocessing of gas (e.g., shipside gas processing), unconventional natural gas, water use and reuse, sustainable or improved extraction, hybrid systems, solar thermal/natural gas, municipal waste/natural gas, small-scale nuclear/natural gas
	Petroleum	Reservoir characterization and engineering, petroleum drilling and recovery technology, enhanced oil-recovery technology, emission control during production and refining, refining, sustainable or improved extraction

Table A.1—Continued

Topic Area	Topic	Subtopic
Energy sources, continued	Biomass	Microalgae, macroalgae
	Waste to energy	Municipal solid waste, facilities for waste water or landfill methane, construction demolition and debris
	Geothermal energy	Geothermal heat pumps
	Hydro, ocean, and tidal energy	Hydrokinetic, thermal gradient
	Hydrogen	Infrastructure, storage, solar-PV electrolysis, solar-assisted electrolysis or solar thermal conversion, nuclear energy for hydrogen
	Nuclear energy	Small-scale hybrid nuclear/gas, radioactive-decay reactors for very small (<10 MW) plants, fission reactors for small (<100 MW) plants, thorium-centered fuel cycle for small plants, cogeneration of desalinated water and nuclear energy, technical dimensions of breeder reactors (e.g., liquid metal), safety, regional and global security implications, waste management, sizing and siting
	Coal	Coal-combustion improvements in pollution reduction, non-GHG pollution control, minimize impact from extraction
Energy storage, delivery, and use	Natural gas use	Gas-turbine development, chemical synthesis (from methane and natural gas liquids), other natural gas–intensive industries and applications (e.g., air conditioning)
	Fuel cells	High- and low-temperature development, mobile and stationary application development, hybrid fuel cells
	Energy storage	Compressed air, batteries, molten-salt thermal storage, supercapacitor and flywheel
	Smart grids	DSM, transmission and distribution technology, high-voltage direct-current lines, integration of variable renewables, advanced meters, grid security, standards for communication and interoperability

Table A.1—Continued

Topic Area	Topic	Subtopic
Energy storage, delivery, and use, continued	Distributed generation of energy	Natural gas (microturbines, fuel cells, advanced diesel), solar (BIPV)
	Solar applications	Solar residential hot water, solar residential cooling and refrigeration, incentives and regulations for residential solar applications
	Alternatively fueled vehicles	Plug-in hybrid, electric, natural gas, LPG (propane), biofuels, hydrogen, retrofit vehicles for natural gas (air quality), fuel-economy standards, transportation-system improvements
	Energy-efficient and green building design	Standards and regulations, energy-efficient technologies
Overarching energy-research areas	Water use and reuse	Solar, natural gas, biofuels
	Development of an energy strategy	Long-term supply and demand assessment, environmental impact and effects, renewable portfolio standards, siting trade-off assessment, strategic and national implications, domestic consumption and export balance
	Capacity building and workforce generation	Domestic and foreign human-capital balance, education and training needs assessment, needs in key discipline areas (e.g., energy policy, engineering analysis)
	Environmental and worker safety	Culture of safety, monitoring and safety
	CCS	Precombustion capture, postcombustion capture, subsurface sequestration
Water sources	Desalination	Desalination technology (e.g., membranes, nanotechnology), alternative energy (e.g., solar desalination), cogeneration of electricity and water

Table A.1—Continued

Topic Area	Topic	Subtopic
Water sources, continued	Groundwater sustainability	Motor and pump efficiency, replenishment technologies (e.g., spreading ponds, injection wells), monitoring technology and planning, saltwater intrusion threats, assessment of groundwater balance, surface groundwater-use optimization, private pumping and well-use monitoring and regulations
	Rainwater capture and harvesting	Excess storm water–capture technology, on-site rainwater-capture technology, capture and harvesting policies
	Wastewater recycling	Technology development; green design of water-treatment plans; microbiology and disinfection; wastewater-use prioritization, optimization, and regulation
	Water imports	Feasibility and cost, economic implications, international implications
	Water quality	Monitoring and assessment technology and regulations, standards
	Distribution-system management	System optimization; energy, leakage, age, health assessment; storage assessment and optimization for potable and treated water
Water use	End-use efficiency	Smart metering, delivery and use devices, urban and household, gray-water systems, low-flow technologies, end-user technology regulations, urban-use regulations, agricultural low-water applications, industrial closed-loop water use
	Water-efficient and green building design	Standards and regulations, water-efficient technologies
	Water demand management	Pricing and regulations, outreach and communication
	Water-use priorities	Optimization of use based on water quality, water-use equity, economic and environmental impact assessment
Overarching water-research areas	IWRP	Model development, conduct IWRP, IWRP policies and regulations
	Governance	Governance options (e.g., unified water authority)

Table A.1—Continued

Topic Area	Topic	Subtopic
Overarching water-research areas, continued	Capacity building and workforce generation	Needs assessment, domestic and foreign human-capital balance, education and training needs assessment, needs in key discipline areas (e.g., engineering, hydrology, manufacturing, policy, chemistry, biology)
	Stakeholder engagement	
	International water issues	International water and precipitation modeling, diplomacy and cross-border collaboration
	Climate change	Assessment of supply and demand, adaptation
Environment	Ambient air, indoor air, drinking water, coastal water, waste, land, biodiversity, food, climate change mitigation and adaptation, environmental economics, occupational health, environmental health	We did not differentiate these topics. Every environmental research topic could include research on sources, fate and transport, exposure, and treatment. Therefore, we determined that further differentiation depends on Qatar's environmental context.

NOTE: LPG = liquefied petroleum gas. IWRP = integrated water resource planning.

Interim Panel Research Topic Recommendations

Expert panels identified 31 promising research topics during their assessment of research for the institute (step 4, selecting priority topics in panels, of the methodology for identifying priority research topics). Experts further categorized these 31 research topics into 11 priority energy topics, four priority water topics, eight priority environment topics, two secondary research topics, and six crosscutting research topics. In this appendix, we present the panel's observations about each priority topic, describe our synthesis of different panels' recommendations, and state how the topic was incorporated into our final recommendations.[1]

Table B.1 lists each of the 23 priority research topics that panels recommended, grouped into energy, water, and environment topics. For each topic, we note whether there was consensus across panels that the topic should be a priority or whether recommendations were mixed across panels. We also note the outcome from the project team's synthesis of recommendations across panels.

Priority Research Topics in Energy (11)

Natural Gas

Panel Observations. Qatar has the third-largest natural gas reserves in the world, approximately 890 trillion cubic feet (TcF). Natu-

---

[1]   Note that, if even one panel recommended a certain topic as a priority, then we have included it in the group of energy, water, and environment priority topics as appropriate.

Table B.1
Table of Panels' Recommended Priority Topics and Outcome from Synthesis

Topic Area	Topic	Panels' Recommendation	Outcome
Energy (11)	Natural gas	Priority	Priority (topic 1)
	Petroleum	Priority	Priority (topic 2)
	CCS	Priority	Priority (topic 3)
	Solar energy	Priority	Priority (topic 4)
	Advanced hydrogen and fuel cells	Mixed	Fuel cells are priority (topic 5); hydrogen is secondary
	Green buildings	Priority	Priority (topic 6)
	Smart grids	Priority	Priority (topic 7)
	Strategic energy planning	Priority	Priority (topic 8)
	ERC	Priority	Included in strategic energy planning (topic 8)
	Efficient water use in energy production	Priority	Included in environmental characterization (topic 13)
	Polymers, aluminums, and plastics	Mixed	Secondary
Water (4)	Desalination	Priority	Priority (topic 9)
	Groundwater sustainability	Priority	Priority (topic 10)
	Water demand management	Priority	Priority (topic 11)
	IWRM	Mixed	Priority (topic 12)
Environment (8)	Environmental characterization	Priority	Priority (topic 13)
	Ecosystem services	Mixed	Subtopic in environmental characterization (topic 13)

Table B.1—Continued

Topic Area	Topic	Panels' Recommendation	Outcome
Environment, continued	Technology procurement	Mixed	Noted in environmental characterization (topic 13)
	Environmental economics	Priority	Subtopic in crosscutting environmental research (topic 14)
	Climate change	Priority	Subtopic in crosscutting environmental research (topic 14)
	Mass–energy balance	Priority	Subtopic in crosscutting environmental research (topic 14)
	Industrial ecology	Priority	Subtopic in crosscutting environmental research (topic 14)
	Closed agriculture systems and advanced aquaculture	Mixed	Secondary

ral gas is the largest source of energy for the country, and Qatar's economy depends largely on its natural gas exports. Given that natural gas is perhaps the most critical resource and critical industry, research in all areas of natural gas should be Qatar's highest priority. This includes characterization, extraction and production, distribution, and use.

Synthesis of Panel Recommendations. Panels agreed that natural gas is a critical resource and a critical industry and should be researched by the institute.

Outcome. As recommended by the panels, we have made research on natural gas one of our priority research areas.

Petroleum

Panel Observations. Petroleum is the second-largest source of energy in Qatar and contributes significantly to its GDP. Therefore, research into petroleum resource characterization, extraction, and processing is also important. It is not of highest priority, and it is not a leadership opportunity, however, because petroleum is not a long-term resource for Qatar and because petroleum research is already a major area of research for other research and industry institutions, including Qatar Petroleum.

Synthesis of Panel Recommendations. Panels agreed that petroleum research should be a priority but that it should be undertaken in consultation with Qatar Petroleum and other institutions already conducting research in this area.

Outcome. As recommended by the panels, we have made research on petroleum one of our priority research areas.

Carbon Capture and Storage

Panel Observations. GHG mitigation is a growing global concern, and, for many countries, natural gas is an important stepping stone from more–carbon-intensive energy sources, such as coal and petroleum, to renewables. Thus, we expect Qatar to have a strong market for its natural gas exports. As emission controls are enacted and made increasingly stringent, however, natural gas might no longer be an appropriate source of energy. CCS—and research in developing CCS—is therefore an important step to ensuring that Qatar's natural gas industry is viable in the future. Given that it has the third-largest natural gas reserves, CCS is more important to Qatar than it might be for other countries, and Qatar has the opportunity to be a leader in this area.

Synthesis of Panel Recommendations. Panels agreed that CCS research should be a priority for the institute and would help Qatar address challenges in natural gas and petroleum.

Outcome. As recommended by the panels, we have made research on CCS one of our priority research areas.

Solar Energy

Panel Observations. Sunlight is one of Qatar's most-abundant yet underutilized resources. Solar energy is a promising option for introducing renewables into Qatar's energy mix and for providing a source of energy when its oil and natural gas resources are depleted.

Synthesis of Panel Recommendations. Panels agreed that, although Qatar could undertake a wide range of research in solar energy, a few areas stand out as particularly important:

- solar thermal technologies, including CSP and ambient air and water conditioning, which are likely to be more cost-effective (in most applications) and stable than PV is
- hybrid solar thermal/natural gas electricity-generation systems, which rely on solar energy when possible and use natural gas for supplemental electricity production
- solar desalination, given Qatar's growing need for water.

Outcome. As recommended by the panels, we have made research on solar energy one our priority research areas. However, solar desalination is recommended as part of a desalination research program rather than as part of a solar energy research program because of its synergies with other desalination research.

Advanced Hydrogen and Fuel Cells

Panel Observations. Qatar could demonstrate how a hydrogen economy might function. Advanced hydrogen production, the use of central and distributed fuel cells, and the research and production of carbon-neutral hydrocarbons and other synthetic fuels would be promising areas of research.

Synthesis of Panel Recommendations. One panel recommended advanced hydrogen and fuel cell research as a priority. Others recommended fuel cells as a secondary topic because it is a high-risk research area and did not recommend advanced hydrogen at all. After further review, we determined that fuel cells are a priority research area given that they could help Qatar use natural gas, a critical resource, more efficiently. However, developing a hydrogen economy is a secondary area of research because it addresses a potential opportunity rather than a critical challenge for Qatar today.

Outcome. Fuel cells are one of the final 14 recommended priority research areas, while hydrogen economy research is a secondary topic in our final recommendations.

Green Buildings

Panel Observations. Green building design can play a key role in addressing Qatar's long-term energy, environment, and water challenges, particularly because there is much new development in the country. Qatar should have high green design standards and regulations, and, as with smart grids, there is significant opportunity for developing expertise in green building policies that could be exported to other countries in the GCC. This includes determining green building priorities, developing standards, and choosing technologies and methods that meet those objectives.

Synthesis of Panel Recommendations. Panels agreed that green building design and adoption research should be a priority.

Outcome. As recommended by the panels, we have made research on green buildings one our priority research areas.

Smart Grids

Panel Observations. Qatar has one of the highest rates of electricity consumption per capita in the world—almost 17,000 kWh per person in 2005. This is in part because electricity is free to Qataris and therefore economic incentives are not aligned with conservation objectives, and in part because it has one of the fastest-growing economies in the world. Demand for electricity is also growing rapidly, and Qatar's electricity production is expected to double between 2008 and 2011 to 9,000 MW. Smart grids, therefore, could play an important role in managing Qatar's electricity sector.

Qatar should strive to become a leader in both smart grid technology and policy research, if efficiency, demand management through pricing and technology, and the use of renewables become key energy goals for Qatar. It might be particularly strategic for Qatar to be a leader in smart grid policies related to implementation pathways and government regulations because smart grid technology research is being undertaken elsewhere and technologies could be imported for use in Qatar, while policy research must be developed for Qatar's social and governmental context and requires little infrastructure investment. It is important to note that, once developed, there might be a large market for Qatar's smart grid policymaking capacity in other GCC nations, which face similar challenges of high and growing electricity demands and might wish to implement their own smart grids.

Synthesis of Panel Recommendations. Panels agreed that smart grid research should be a priority because smart grids can address many of the challenges that Qatar's electricity sector faces.

Outcome. As recommended by the panels, we have made research on smart grids one of our priority research areas.

Strategic Energy Planning

Panel Observations. It is extremely important for Qatar to develop an energy strategy that balances its long-term goals, resources, and constraints and to continue to refine this strategy over time as goals, resources, and constraints change. In addition to shaping Qatar's energy sector, this strategy will identify future research priorities for the country. It is also important for Qatar to have the capacity to develop an energy portfolio domestically. A key subarea of research is determining the most-efficient domestic uses of Qatar's natural gas (i.e., applications, such as fuel cells and hybrid solar thermal/natural gas systems) that use the least gas per unit of service provided and balancing this with exports to maximize the potential of Qatar's natural gas reserves.

Synthesis of Panel Recommendations. Panels agreed that research to help develop a long-term energy strategy is important for existing and future critical energy resources and should be a priority.

Outcome. As recommended by the panels, we have made research on strategic energy planning one of our priority research areas.

Energy Resource Characterization

Panel Observations. Although much might be known about Qatar's oil and natural gas reserves, it is important to also characterize other energy resources, including solar, wind, geothermal, and ocean or tidal. The institute could serve a research function in Qatar similar to that of the U.S. Geological Survey in the United States, providing scientific information about Qatar's natural resources and environment.

Synthesis of Panel Recommendations. Panels agreed that a long-term ERC is important in developing future energy sources.

Outcome. We recommend ERC as a subtopic in research to develop Qatar's long-term energy strategy (topic 8).

Efficient Water Use in Energy Production

Panel Observations. Water is a critical resource in Qatar: Qatar lacks surface water, its groundwater is being rapidly depleted, and the opportunity cost of desalinating water is expected to grow. Yet Qatar's two largest sources of energy—natural gas and petroleum—require significant amounts of water during production. So, too, would new sources of energy that Qatar might consider in the future, such as solar energy.

Synthesis of Panel Recommendations. Panels agreed that the efficient use of water in energy production, including reduction in initial water use, options for water reuse, and options for water sources, such as seawater, should be high-priority research topics. However, later research revealed that Qatar uses seawater for once-through cooling. After further discussions with experts, we determined that the main hazard from this particular use of water is the negative effect on marine ecosystems from water intake and thermal and mineral pollution.

Outcome. We recommend efficient water use and safe water intake as subtopics under topic 14, environmental characterization.

Polymers, Aluminums, and Plastics

Panel Observations. Given the abundance of energy and fossil fuel feedstocks, Qatar could research and develop more-efficient methods of producing polymers, aluminum, and plastics and could develop industries around this research.

Synthesis of Panel Recommendations. One panel recommended polymers, aluminums, and plastics research as a priority, while others did not recommend it. After further review, we determined that this is a secondary area of research because it addresses a potential opportunity rather than a critical challenge for Qatar today.

Outcome. Polymers, aluminums, and plastics research is a secondary topic in our final recommendations.

Priority Research Topics in Water (4)

Desalination

Panel Observations. Desalination is of critical importance as a source for water in the country. The opportunity cost of desalination is expected to increase as and when Qatar's oil and natural gas resources are depleted, so innovation in desalination technology should be a top priority. Specifically, solar thermal seems promising because of Qatar's climate, and Qatar could become a leader in this area. Other desalination technology research might also be appropriate, and a more in-depth assessment of those technologies would be warranted in order to expand on this recommendation. The other area of research is the cogeneration of electricity and desalinated water, which is both a policy research area (siting) and a technology research area.

Synthesis of Panel Recommendations. Panels agreed that desalination research would address critical water-security challenges in Qatar and should be a priority research area.

Outcome. As recommended by the panels, we have made research on desalination one of our priority research areas.

Groundwater Sustainability

Panel Observations. Qatar's groundwater is deteriorating rapidly due to unchecked and inefficient use, limited recharge, and saltwater intrusion. If Qatar does not address this issue, its groundwater could be permanently depleted, and Qatar would be dependent solely on desalination for water. Qatar should instead be a leader in groundwater sustainability in arid climates and undertake a comprehensive research agenda that includes developing methods of improving groundwater recharge (e.g., via artificial recharge), reducing evaporation (e.g., through runoff capture), and remediating groundwater. This would include research in technology, policy, and resource management. Precipitation and temperature changes due to climate change make groundwater sustainability a moving target, as is understanding the implications that climate change has for long-term groundwater sustainability in the Gulf region.

Synthesis of Panel Recommendations. Panels agreed that groundwater is a critical national resource, and research should be a priority in order to address water-security challenges in Qatar.

Outcome. As recommended by the panels, we have made research on groundwater sustainability one of our priority research areas.

Water Demand Management

Panel Observations. Qatar has undertaken a variety of supply-side projects and research. However, demand-side research seems lacking, and water consumption appears to be a critical challenge. Qataris do not have to pay for their water use, there is not yet a culture of conservation, and the delivery and use systems are inefficient (according to some sources). Therefore, Qatar should also undertake a comprehensive research agenda to reduce the demand for water, mainly through policy research. This includes questions of how to introduce and effectively apply pricing and regulations to have less water-intensive applications, questions of how to create outreach programs (e.g., in primary and secondary schools) so that future generations appropriately value and conserve water, policies that encourage the use of (and make available) efficient water and delivery devices, and policies to assess and improve the water-distribution infrastructure. There is also scope for developing efficient-use technologies, notably in agriculture, in which there appears to be significant waste.

Synthesis of Panel Recommendations. Panels agreed that demand-management research would address critical water-security challenges in Qatar.

Outcome. As recommended by the panels, we have made research on water demand management one of our priority research areas.

Integrated Water Resource Management

Panel Observations. Qatar cannot address its water problems in a piecemeal fashion: Water production and consumption are intimately linked to many sectors, including households, agriculture, and industry. Qatar must institutionalize ongoing IWRM that takes into account the relationships between demand from different sectors, different supply options, policies, and regulations.

Synthesis. One panel recommended that IWRM research be a priority to help address critical water-security challenges in Qatar. Another panel did not mention IWRM specifically but did mention related water-security issues. After further discussion with experts, we determined that, like energy strategy development, IWRM is key to long-term water planning.

Outcome. IWRM is one of the final 14 recommended priority research areas.

Priority Research Topics in Environment (8)

Environmental Characterization

Panel Recommendation. Qatar's research should, first and foremost, focus on characterizing its environment, given that there is a great deal of uncertainty about the health of Qatar's coastal waters, fisheries, deserts, ambient and indoor air, and other natural resources. This includes (1) determining baseline environmental health, (2) monitoring pollutants, and (3) developing models of fate and transport. Once Qatar has characterized its environmental concerns, it should acquire (rather than develop) pollution-control solutions, given that (1) such solutions have been developed elsewhere and can be purchased and (2) Qatar is not uniquely positioned to be a leader in developing new solutions. In many of the research areas, such as air quality and coastal waters, regional modeling and analysis are required, given that pollution and ecosystems cross international boundaries.

Synthesis. Panels agreed that environmental characterization is a key first step in addressing Qatar's environmental challenges.

Outcome. As recommended by the panels, we have made research on environmental characterization one of our priority research areas.

Ecosystem Services

Panel Recommendation. As desertification continues in the region, the institute should seek to understand the ecosystem services provided by the local desert and surrounding gulf and the impact that climate change has on these services. Ecosystem services can be categorized as provisioning (food, water, energy), regulating (climate, wastes, purification), supporting (nutrient and seed dispersal), and cultural services.

Synthesis. Panels agreed that assessing ecosystem services would help to determine the value of Qatar's ecosystem.

Outcome. We include ecosystem services as part of the environmental characterization research program, which is one of the final 14 recommended priority research areas.

Technology Procurement

Panel Observations. Procurement policies can determine the eligibility, appropriateness, and effectiveness of funded projects and programs. For example, policies that allow no-bid contracting might result in funding decisions that differ from those arising out of policies that require competitive bidding. In the context of environment issues, Qatar's procurement policies could affect how and which pollution-control and environmental solutions are acquired.

Synthesis Across Panels. One panel noted that the institute could help Qatar assess and refine its procurement policies in order to improve Qatar's ability to implement solutions to its environmental problems. Other panels did not mention this.

Outcome. We note that technology procurement might be part of next steps addressing environmental challenges identified through the process of environmental characterization, which is one of the final 14 recommended priority research areas.

Environmental Economics

Panel Recommendation. Qatar needs to conduct environmental economics research so that the value of environmental risks can be determined and so that policies—particularly market-based approaches—can be developed, evaluated, and compared, to address those risks.

Synthesis. Panels agreed that environmental economics could affect how Qatar determines the value of environmental risks.

Outcome. We include environmental economics as part of a research program on crosscutting environmental research topics, which is one of the final 14 recommended priority research areas.

Climate Change

Panel Recommendation. Climate change is expected to make Qatar more arid, and this, in turn, could have effects on biodiversity, coastal waters and fisheries, desert ecosystems, other environmental concerns, and human health. Qatar should research the anticipated effects that climate change could have on its environment.

Synthesis. Panels agreed that the institute should research how climate change could affect Qatar's environment.

Outcome. We include climate change as part of a research program on crosscutting environmental research topics, which is one of the final 14 recommended priority research areas.

Mass–Energy Balance

Panel Recommendation. The institute should also develop a complete mass–energy balance of the country if this has not already been done. It is critical that Qatar understand the composition of its resource use. By initiating a characterization of energy and materials going into the economy, examining their pathways, and determining losses and wastes, Qatar can then optimize for efficiency. Development of economic and environmental input-output accounts for Qatar would be important first steps.

Synthesis. Panels agreed that a mass–energy balance would help Qatar increase its industrial efficiency and protect its environment.

Outcome. We include mass–energy balance as part of a research program on crosscutting environmental research topics, which is one of the final 14 recommended priority research areas.

Industrial Ecology

Panel Recommendation. Industrial-ecology research and the development of zero-waste industrial parks could help Qatar reduce the environmental impact of its hydrocarbon and other industries by optimizing industrial processes, energy, and waste.

Synthesis of Panel Recommendations. Panels agreed that industrial-ecology research is important because of its implications for pollution and waste.

Outcome. We include industrial ecology as part of a research program on crosscutting environmental research topics, which is one of the final 14 recommended priority research areas.

Closed Agriculture Systems and Advanced Aquaculture

Panel Observations. The smallness of Qatar's agriculture sector suggests that food security is a key concern. Qatar might be able to address its food security and environmental concerns by researching and developing completely closed systems of plant and animal agriculture, as well as advanced aquaculture.

Synthesis of Panel Recommendations. One panel recommended this area of research as a priority, while others did not recommend it. After further review, we determined that, given that Qatar has a small agriculture sector, this is a potential opportunity for the future rather than a critical challenge for Qatar today.

Outcome. We have included this as a secondary topic in our final recommendations.

Secondary Research Topics (2)

Marine-Algae Biofuels

Panel Observations. Algae biofuels might be high-risk, high-reward research. Although use of marine algae for biofuel has not yet been achieved, it would use Qatar's ocean resources in a new way, and it would produce oil, for which there is already a market.

Outcome. This is noted as a secondary research topic in the final recommendations.

Distributed Energy Generation

Panel Observations. Distributed energy generation includes developing microturbines, BIPVs, and similar technologies and would be valuable if Qatar is interested in providing distributed electricity to those outside of Doha.

Outcome. This is largely covered under solar, smart grids, and green buildings, so we have not retained it in the final recommendations as a separate secondary topic of research.

Crosscutting Research Topics (6)

Occupational Health Research

Panel Observations. Occupational health research addresses concerns about how pollutants, climate change, and other environmental concerns can affect worker health. This research could be undertaken in partnership with a health research institute.

Outcome. This is noted as a crosscutting topic in the final recommendations.

Food Safety

Panel Observations. Food-safety research addresses concerns about pesticides in food. This research could be undertaken in partnership with a health research institute.

Outcome. This is noted as a crosscutting topic in the final recommendations.

Environmental Health

Panel Observations. Environmental health addresses the environment's effects on the health of the general population (e.g., respiratory illness due to poor ambient air quality, building sickness due to poor indoor air quality). This research could be undertaken in partnership with a health research institute.

Outcome. This is noted as a crosscutting topic in the final recommendations.

Ultrapure Water and Industrial Water Use

Panel Observations. Ultrapure water and industrial water use could be a valuable area of cross-collaboration between Qatar's computing institute and the energy and environment research institute. For example, if Qatar's computing and information technology fabrication research requires ultrapure water, research on optimal production methods could be conducted.

Outcome. This is noted as a crosscutting topic in the final recommendations.

Materials Research

Panel Recommendations. The institute should undertake materials research to further its technology development goals across the energy and environment, computing, and health institutes.

Outcome. This is noted as a crosscutting topic in the final recommendations.

Social-Science Research

Panel Observations. Each of QF's institutes seeks to conduct research in a sociotechnical area, so a social-science research track that addresses social and decisionmaking issues in and across each area is important. This can include quantitative and qualitative policy analysis, an understanding of the externalities of local decisions, and innovation metric research and other methods to characterize the local competitive advantage.

Outcome. This is noted as a crosscutting topic in the final recommendations.

Profiles of Institutions in the Gulf Cooperation Council

The institutions listed in Table C.1 are among the most-prominent ones engaged in energy and environment research in the GCC. The following profiles are based on our surveys, literature reviews, and site visits and provide information on each institution's history, organization, research priorities, and current or planned activities. The institutions are organized by country. More-comprehensive information on the types of research in which each engages can be found in the GCC Energy and Environment Research Database, which we have provided to QF's vice president for research.

Table C.1
Environment and Energy Research Institutions in the Gulf Cooperation Council

Country	Institution
Qatar	QNRF, QSTP, QU, Texas A&M University at Qatar, industry and other private-sector research
UAE	UAEU, Masdar Institute of Science and Technology
Kuwait	KISR
Oman	SQU, MEDRC
Bahrain	Bahrain Institute for Studies and Research
KSA	KACST, KAUST, KFUPM

Qatar

Qatar's research capabilities in energy and the environment are still nascent; most institutions have been established in only the past decade. Basic and some applied research are being undertaken at academic institutions, such as Qatar University and the Education City universities. Many of the companies operating in Qatar, particularly in the oil and gas sectors, have historically conducted their research and development activities in laboratories outside Qatar. However, with the opening of the free-trade zone at QSTP, the research and development presence in Qatar is growing. Several companies in QSTP are engaged in commercializing technologies. There is much potential for collaboration and synergies between institutions in Qatar and those in the United States and Europe. However, regional partnerships in energy and environment are limited at the moment.

Qatar National Research Fund

QF established QNRF in 2006 as a national resource for research. QNRF reviews and competitively funds selected research with the aim of helping to advance knowledge in Qatar and globally. Funding is provided for a wide range of subjects, including engineering and technology, physical and life sciences, medicine, humanities, social sciences, and the arts.[1] QNRF conducts three funding cycles annually. Each cycle can grant up to $86 million in grants ranging in amount from $20,000 to $350,000. Financial opportunities can be awarded to researchers with a wide range of qualifications, from students to professionals, in the private, public, and academic sectors.[2]

In addition to providing funding, QNRF is involved in other research activities, ranging from conducting national research surveys to joint ventures designed to commercialize research. QNRF programs and projects include the following:[3]

---

[1] QNRF, undated.

[2] QNRF, undated.

[3] QNRF, undated.

- Undergraduate Research Experience Program (UREP)
- National Priorities Research Program (NPRP)
- workshops, conferences, and short courses
- joint ventures in innovative enterprises
- distinguished fellowships
- biennial national research survey
- outreach service to Qatari institutions for applied research.

A governing board of six members sets the fund's overall direction, policy, and oversight. Stakeholders include QNRF staff and members of an advisory council comprised of academic and business leaders from Qatar.[4] The NPRP and the UREP are the two main funding priorities for QNRF.[5] Awards are granted for research pertaining to the needs and aspirations of Qatar. Research areas of priority to Qatar include the following:

- computer and information sciences (e.g., geographic information system [GIS], cybersecurity)
- earth and related environmental sciences (e.g., air quality, water distribution)
- chemical sciences (e.g., CO_2 capture)
- biological sciences (e.g., coastal ecosystems)
- electrical engineering (e.g., robotics)
- nanotechnology (e.g., new materials)
- mechanical engineering (e.g., airport efficiency, GTL processing)
- materials engineering (e.g., recycling, remediation)
- medical and health sciences (e.g., diabetes, obesity)
- agricultural sciences (e.g., desertification)
- social sciences (e.g., economic diversification)
- humanities (e.g., Shari'a).

---

[4]   Greenfield et al., 2008.

[5]   Greenfield et al., 2008.

Only institutions within Qatar are eligible to submit proposals for QNRF funding, and the co–principal investigator must be based in Qatar and affiliated with the submitting institution.[6]

Qatar Science and Technology Park

QSTP serves as both an innovation hub and applied research center facilitating and housing several technology-centered commercial companies. It falls under the umbrella of QF and is located in QF's Education City, enabling industry–university collaboration with such institutions as Texas A&M University at Qatar, which has established a research center to study the utilization of natural resources and environmental sustainability, and the clinical medical research program at Weill Cornell Medical College. It operates in a free-trade zone, which helps attract foreign companies.[7] It offers several support and business-incubation programs to help companies develop and commercialize their research. QSTP research receives a majority of its funding from QNRF.

The research conducted in QSTP varies by company and is mainly focused on applications with significant commercial potential. Some of the key research areas in energy and environment include the following:[8]

- Water: water quality, distillation, and management (General Electric's and ConocoPhillips joint Global Water Sustainability

---

[6] QNRF, undated.

[7] Companies conducting research at QSTP as of December 2010 include AES International Consultants, BQDRI, Chevron Qatar Energy Technology, Cisco Systems, ConocoPhillips, deltaDOT QSTP-LLC, EADS, ExxonMobil, Fuego Digital Media, GE, GreenGulf, Gulf Bridge International, Hydro, iHorizons, Institut de Soudure, MEEZA, Microsoft, Qatar Petroleum, Qatar Robotic Surgery Centre, Qatar University Wireless Center, Rolls-Royce, Shell, Tata Consulting Engineers, Total, Transport Research Laboratory, Virgin Health Bank, and Williams Grand Prix Engineering Limited.

[8] There are also health and information technology (IT) projects under way that are unrelated to energy and environment but that might be important for QF's planning for health and IT centers of excellence.

Center), combined desalination and power generation and novel water treatment
- Green building: modular energy-efficient housing
- Energy: CCS (Qatar Petroleum, Shell), reservoir engineering (Qatar Petroleum, Shell), renewable energy sources (GreenGulf), synthetic aviation fuel and high-value oil and gas derivatives.

As a free-trade zone, QSTP's facilities offer several benefits, including the ability to hire expatriate employees, no taxes, duty-free imports of goods and services, and unrestricted repatriation of capital and profits.

In addition to the businesses that reside in the QSTP facilities, QSTP has partnerships with Virginia Commonwealth University in Qatar (VCUQatar), Airbus, University of Sheffield, Imperial College London, Germany's Space Agency (Forschungszentrum der Bundesrepublik Deutschland für Luft- und Raumfahrt, or DLR), Biogem Italy, Excalibur Fund Managers Limited, Virgin Group, Qatar Airways, Qatar University, and Aspetar.

Qatar University

Qatar University was founded in 1973 as a national institution of higher learning with four colleges and has since expanded to ten.[9] The Office of Academic Research (OAR), which was established in 2007, manages research activities for the university. Applied research relevant to energy and environment is conducted in the Gas Processing Center and the Materials Technology Unit. These specialized centers support applied research through collaboration with industry and government partners. Funding for research in these centers comes mainly from NPRP and UREP grants. Research is also conducted through private-sector funding and consultancies. About 66 percent of the total grants that QU receives are spent on research on energy and the environment.[10]

---

[9] At present, the colleges are the Colleges of Arts and Sciences, Business and Economics, Education, Engineering, Law, Pharmacy, Sharia and Islamic Studies, the Foundation Program, the Sport Science Program, and the Honors program.

[10] Al-Derham, 2010.

The Materials Research Group takes a multidisciplinary approach, drawing on faculty and students from various departments across the university. It focuses on projects in the following areas, with applications for the oil and gas industries, green buildings, and waste management:[11]

- materials processing and characterization
- electrochemistry and corrosion
- composites
- recycling and biodegradable materials
- computational studies.

The Gas Processing Center was established in 2007 through QU's engineering department to support applied research for the gas industry in Qatar. The department's multidisciplinary research program is focused on gas processing and environmental impacts. It has four teams:[12]

- Materials (through the Materials Research Group): CCS, hydrogen production, storage and transportation, materials performance, erosion, corrosion and wear, and new materials and processes
- Environment: water and wastewater treatment, solid-waste management, and soil bioremediation
- Energy: process optimization, separation processes and technologies, natural gas processing, and CCS
- Controls and instrumentation: wireless sensing and communication, power controls and management, condition monitoring, and process control and optimization.

Areas of interest for development and expansion in Qatar include biofuels, desertification, marine issues, air-quality monitoring and

---

[11] Qatar University, 2011b.

[12] Gas Processing Center, undated (a).

management, and green buildings.[13] QU is also exploring a Ph.D. program to support research at the Gas Processing Center and the possibility of developing a nuclear research laboratory. One challenge QU faces is limited human resources for research. Collaboration with institutions in Qatar and outside the GCC is fairly common, but there are few collaborative activities between QU and institutions within the GCC. About 14 percent of the collaborating partners for NPRP-funded projects in the Materials Technology Unit are other institutions in Qatar. About 50 percent of the partners are in North America. Only 4 percent of the collaborating institutions are in the MENA region (Jordan and Saudi Arabia).[14]

Texas A&M University at Qatar

Texas A&M University opened its Qatar branch in 2003 in Education City. It offers bachelor of science degrees in mechanical, electrical and computer, and petroleum and chemical engineering. In 2011, the university will begin offering a master of science degree in chemical engineering. Faculty and students are involved in research and receive funding through QNRF and private consultancies, particularly from the oil and gas industries. Specialized laboratories for this research include the following:

- Fuel Characterization Laboratory
- Fuel Cell Laboratory
- Multi Scale Thermo Fluids Laboratory
- Gas Condensate and Wettability Research Laboratory
- Acid Stimulation Laboratory
- Wireless Communication Lab
- Materials Characterization Laboratory
- Power Laboratory.

---

[13] QU won a grant to collaborate with Qatar Airways and QSTP on the development of aviation biofuels. Marine issues would focus on monitoring and management of concentrated brine and chlorine in coastal waters.

[14] Qatar University, 2011c.

In 2010, the university began the Qatar Sustainable Water and Energy Utilization Initiative (QWE), an effort designed to encourage a multidisciplinary approach to water and energy issues.

Research priority areas include "more effective oil and gas production, innovative methods to convert Qatar's abundant supply of natural gas to ultra-clean transportation fuels, faster development of alternative energy sources, novel methods for hazardous waste water treatment and enhancement of chemical process safety."[15] Specific focus areas include the following:

- Energy: Advanced communications for gas exploration, drilling and production, GTL technologies and uses, synthetic fuels, use of nanotechnology in heat transfer, and alternative energy sources
- Solar: Use of solar energy to produce hydrogen from methane
- Water: Novel desalination techniques and novel methods to prevent toxic compounds from contaminating coastal water near industrial sites.

The university collaborates with other public and private institutions in Qatar, including QU, Ministry of Environment, Ministry of Energy and Industry, ExxonMobil, Shell, and Sasol. The university works in close cooperation with its campus in the United States, which also has relationships with other academic institutions around the globe. Researchers receive requests for collaboration from many institutions both in the region and globally. However, several researchers report that project portfolios are saturated.[16]

Industry and Other Private-Sector Research

Although not an institution per se, industry carries out important research, so we profile it here. The oil and gas industries are among the major supporters of research on energy and the environment in Qatar. The RQPI team held discussions with two companies conducting research at QSTP, one of which is a major energy company,

---

[15] Texas A&M University at Qatar, undated.

[16] Informal discussions with university researchers.

ConocoPhillips and GE's Global Water Sustainability Center (GWSC), and the other is GreenGulf.

The GWSC was inaugurated April 2009 with the goal of examining ways to treat and reuse by-product water from oil production and refining operations. The GWSC is a 50/50 collaborative effort between ConocoPhillips and GE. Although the majority of project work is focused on water issues in the petroleum and petrochemical sectors, additional emphasis is placed on water use in the industrial sector and water conservation and municipal issues in Qatar.

The research focus of the GWSC is addressing the following sustainability issues:[17]

- reducing freshwater consumption and promoting water reuse
- lowering water-related operational costs
- minimizing impact on the environment.

The GWSC's QSTP facility includes office space and two research labs with equipment for chromatography, spectroscopy, wet chemical analyses, and application testing. The labs will eventually accommodate approximately 15 researchers. However, at the time of the interview, the laboratory facilities were neither staffed nor operational.

GreenGulf focuses on the development, management, and marketing of renewable energy in the MENA region.[18] It is the first company of its kind to be established in Qatar. GreenGulf has many lines of business: acting as adviser for clean technology and renewable energy, a policy and implementation developer and consultant, and a business promoter.[19] The company has vast experience in developing greenfield projects and strong partnerships across the clean-technology sector.[20] Its main commitment lies in the development of energy efficiency and

---

[17] GWSC, 2011.

[18] GreenGulf, undated.

[19] GreenGulf, undated.

[20] GreenGulf, undated.

renewable energy to address environmental and economic issues, such as climate change and the growing need for clean power generation.[21]

GreenGulf also maintains a research focus on renewable energy technology, policy, and economics. The company has a five-year research plan with QSTP, and its staff members are housed in shared office space with Chevron. It is currently undertaking a joint project with Chevron to conduct a solar energy study for Qatar. This project will use space allocated by QSTP to test new solar technologies. The data collected by GreenGulf will help to identify the most-promising technologies for Qatar's specific climate and terrain conditions.[22]

United Arab Emirates

The UAE, like Qatar, is pursuing sustainable development. A variety of new research initiatives have been undertaken in recent years. The two institutions in the UAE that we profile are the most established (UAEU) and the newest (Masdar Institute of Science and Technology). Both academic institutions engage in applied energy and environment research. Other institutions engaged in this type of research in the UAE include Ajman University of Science and Technology (AUST), CERT, the Gulf Research Center, the Rochester Institute of Technology, and the Energy and Resources Institute (TERI) Gulf Centre.[23]

United Arab Emirates University

UAEU, located in Al Ain, was the first university to be established in the UAE. It was founded in 1976 with four colleges. In 2010, UAEU had nine colleges, more than 12,000 students, and more than 700 faculty members.[24] The institution offers undergraduate programs, postgrad-

---

[21] GreenGulf, undated.

[22] Al Kuwari, 2010.

[23] We chose not to profile these institutions in this chapter because their applied work in energy and environment is fairly small in comparison to other institutions. More information on these institutions can be found in our Environment and Energy Database.

[24] UAEU, 2011.

uate courses, research facilities, and continuing-education programs. UAEU is primarily a teaching university. However, faculty members are expected to spend at least one-third of their time on research. UAEU's vision is to "become a world-class center for applied research, national and international outreach, innovation and outcome-based learning."[25]

Research is funded both internally and externally with resources from both the public and private sectors. Faculty members are typically responsible for identifying and securing sources of funding for their research. Seed funding is available for about 130 grants of $50,000 annually. Since 1998, UAEU faculty members have received research grants from private companies. Faculty members now interact with about 200 companies. Internal funding awards average about 30–40 per year. An Abu Dhabi research foundation is in the process of being created. The foundation will provide $20 million to $30 million annually for research projects at universities in the UAE.[26]

Research on energy and the environment does not make up a large portion of the total research being conducted at UAEU, although it is an area of interest among many of the researchers. The colleges currently engaged in energy and environment research as part of their research portfolio include the Faculty of Science, the Faculty of Engineering, and the Faculty of Information Technology. The engineering faculty has about 95 members with 40–45 ongoing research projects and about 13 related to energy.[27] Key areas of environment and energy research include the following:

- Water: Improving the efficiency of wastewater treatments, nanotechnology for desalination and extracting contaminants from groundwater, and use of electronic sensors and information technology to manage and reduce system leaks
- Solar power: Efficient capture of solar energy and hybrid solar/wind systems

---

[25] UAEU, 2011.

[26] UAEU staff and researchers, 2010.

[27] Almehaideb, 2010.

- Energy: Improving carbon capture, recovery, and sequestration; sensor computing for continuous control in oil and gas fields, smart grid systems, biofuels, and nanotechnology applications for clean energy[28]
- Sustainable building: Green building materials, reduction of water use in construction, sustainable building design, and natural ventilation and cooling for buildings
- Environment: Air-quality monitoring and testing.

Collaboration within GCC countries and even between emirates is a challenge for UAEU. It is often easier to collaborate with universities in Europe or the United States. The exception is collaboration with SQU in Oman: UAEU has maintained this partnership through a formal arrangement for six years now and has funding set aside on an annual basis for collaborative projects. The two universities have tried to coordinate some projects with Qatar University, but funding restrictions and issues with intellectual property (IP) rights make this difficult.[29] QNRF funding for projects requires that one of the principal investigators (PIs) be resident in Qatar; in Saudi or Kuwait, the requirement for the proportion of the work done in country can be as high as 70 percent. A researcher in Qatar is limited to being the PI on no more than six projects. Researchers often quickly meet their limit, making it difficult to initiate collaborative projects.[30]

Masdar Institute of Science and Technology

Masdar Institute of Science and Technology was established in September 2009 as a nonprofit, research-oriented, graduate-only institute located in Masdar City, Abu Dhabi. To take advantage of existing synergies in energy and environmental research, the institute is associated with the Masdar Group of the Abu Dhabi Mubdala Development Company, which includes three business units and an investment

---

[28] UAEU's IT faculty has a grid-computing laboratory, which is unique to the region.

[29] UAEU staff and researchers, 2010.

[30] UAEU staff and researchers, 2010.

arm for the development and commercialization of renewable energy solutions:

- Masdar Carbon develops energy efficiency and clean fossil fuel projects.
- Masdar City is a clean technology site being developed near Abu Dhabi to be powered entirely by renewable energy. It is the world's first carbon-neutral city and is intended to be a living laboratory for researchers at the institute.
- Masdar Power builds and invests in utility-scale renewable energy power projects.
- Masdar Venture Capital manages the Masdar Clean Tech Funds, a series of diversified venture-capital investment vehicles, and will assist in commercializing applied research.[31]

Masdar Institute was developed in conjunction with MIT, which provides continued curriculum and faculty-recruitment support for the seven masters degree programs currently offered and the planned inter-disciplinary doctoral degree program that will be launched in 2011.[32]

At least 50 percent of student and faculty time is dedicated to research.[33] An interdisciplinary research approach that incorporates technology, systems, and policy is a core tenet of Masdar Institute's operations. The research spectrum spans basic to applied research. Unlike some of the other educational institutions in the GCC that have much-broader mandates, Masdar Institute offers only graduate education and focuses almost entirely on applied energy and environment research. Specific focus areas include these:

---

[31] Masdar Institute, undated (a).

[32] Current masters of science programs include computing and information science, electrical-power engineering, engineering systems and management, materials science and engineering, mechanical engineering, microsystems engineering, and water and environment planning. In 2011, Masdar will add masters of science degrees in green chemical engineering and smart infrastructures.

[33] Masdar Institute, undated (b).

- Sustainable building: Using access to Masdar City to research sustainable construction and operation of buildings, green building materials, natural ventilation and cooling, improved insulation, and reduction of water use in construction
- Solar power: Improving photocell and PV performance through advanced materials
- Water: Improving wastewater treatment, efficient water management, water quality, possibilities for seawater irrigation, and use of nanotechnology in desalination
- Energy: Technologies for CCS and policies for carbon sales (in coordination with Masdar Carbon), production of biofuels for aviation, integration of renewable energy into conventional power systems, and developing smart grids
- Policy research: Environment and sustainable development policies.

Masdar Institute does not have any regional partners in the GCC but has fellowship partnerships with institutions in Europe, East Asia, and North America and research projects with multinational corporations, such as Boeing and Honeywell UOP.[34] Masdar Institute is a very young institution. Challenges include building an international reputation for excellence in research, as well as recruiting and retaining high-quality researchers and faculty.

Kuwait

Kuwait has a long history of energy research since the founding of KISR in 1967. As is the case with other GCC countries, the impetus for applied research in energy and environment initially came from exploration and exploitation of petroleum resources. Both internal pressures from economic and population growth and international

---

[34] The fellowship partnerships are with Imperial College London in the United Kingdom, Rheinisch-Westfälische Technische Hochschule Aachen in Germany, Tokyo Institute of Technology in Japan, University of Waterloo in Canada, and the University of Central Florida in the United States.

pressures for environmental reforms have spurred research in sustainable technologies and energy diversification. We chose to profile KISR, which is a unique institution in the GCC as a mature, stand-alone research facility.

Kuwait Institute for Scientific Research

Established in 1967, KISR is the longest-running applied research center in the GCC. It was initially founded by a Japanese firm, the Arabian Oil Company Limited, in fulfillment of its obligations under the oil concession agreement with the government of Kuwait.[35] The institute was originally established to carry out applied scientific research in three fields: petroleum, desert agriculture, and marine biology. The present mission of the institute is

> [p]romoting scientific and applied research, particularly in matters related to industry, natural and food resources and other primary constituents of national economy, in an endeavor to serve the goals of economic, technological and scientific development, and to advise the government on scientific matters and on scientific policy issues.[36]

KISR is a public institution. Its director reports directly to the Minister of Education, but KISR is not considered to be under that ministry. A board of trustees provides oversight for the institute, and the director-general is in charge of daily operations.

The institute undertakes research and technical consultations for both government and private institutions in Kuwait and in the region. Funding has not been a challenge for the institute. Annual funding for the institution is about KD 56 million (about $190 million). About 80 percent of the funding comes from the Kuwaiti government, and the remaining 20 percent is from private companies and other institutions.[37]

---

[35] KISR, undated.

[36] KISR, undated.

[37] Al-Mutairi, 2010.

KISR has more than 70 specialized research laboratories housed in its main research center in Kuwait City and several satellite centers. KISR also manages a network of 80 monitoring stations for radioactivity and radioactive materials. The institute employs approximately 844 staff, 79 percent of whom are Kuwaiti.[38] The institute faces some challenges in finding and hiring qualified research staff due to high demand in the region.

KISR's key research areas are environment and urban development, food resources and marine science, petroleum research and studies, techno economics, and water resources. Apart from research, the institute also conducts training courses in oil testing, management, and other energy programs. Specific research areas include the following:

- Environment: Monitoring marine and air pollutants, power plant intakes and outtakes, radiation and radioactive materials, strategies for combating desertification, coastal management, and improving environmental rehabilitation
- Energy: Policies for CO_2 trading, technologies for improving the efficiency of oil recovery, use of nanotechnology for energy applications, improving thermal-energy storage, technologies or strategies for reducing capital investment in renewable energy projects, and country-wide wind mapping to determine ideal locations for wind farms
- Sustainable building: Development of sustainable building materials and aggregates, improving energy efficiency in buildings[39]
- Solar: Improving solar infrastructure and technologies to overcome limitations and shortcomings of current solar collection systems
- Water: Developing novel desalination technologies and improving existing technologies and new technologies and efficiency gains for wastewater management.

---

[38] Al-Mutairi, 2010.

[39] Kuwait has limited domestic resources for conventional building materials.

KISR is part of the worldwide seismic network and has stations in Oman and Dubai. It also works with the International Energy Agency for research and monitoring of radioactive materials. KISR has collaborated with the GCC to develop the Gulf Electronic Scientific Research (GESR) database.[40] Little collaboration occurs between KISR and GCC institutions, but one researcher is currently engaged in a collaborative effort with a researcher at Qatar University.[41]

KISR identifies its strength as providing an established, stable, and sustainable research environment. It stands out in the region for its development of detailed wind mapping and radiological surveys for Kuwait. This database and expertise could be helpful to other Gulf countries that wish to identify opportunities for alternative energy. KISR's strategic plan for the next five years includes a more interdisciplinary approach to research by appointing cross-sectional monitors. KISR plans to develop a more robust commercialization and marketing capability, including a technology-transfer arm.[42]

Oman

Limited freshwater resources, a coastline of about 1,700 kilometers, and a history of fishing and maritime trade have shaped institutional research priorities in Oman. Institutions engaged in energy and environment research tend to focus on water resources and marine issues. Oman's biodiversity relative to other Gulf countries make it a regional leader in environmental research. Oman's economy depends heavily on the oil and gas sectors, as do those of other GCC nations, so research institutions are active in developing applied technologies for the petroleum industry. In June 2005, the Research Council was established by royal decree to provide strategic direction for research and to act as a national public fund for research in Oman. The Research Council has

---

[40] We have attempted to gain access to this database but have been unable to get permission from the GCC.

[41] Group interview with KISR researchers.

[42] Al-Mutairi, 2010.

helped to establish priority research areas in the energy, industry, biological, and environmental-resource sectors.

The two institutions that we selected to profile in Oman are SQU and its research centers and MEDRC, a funding organization for water-research activities in the MENA region.

Sultan Qaboos University

SQU was established in 1986 and is the most prominent academic institution in Oman. Research is an important function of the university. Faculty promotion is closely tied to research performance. Research is conducted in the university's nine colleges and seven research centers. The departments engaged in energy and environmental research include agriculture and marine sciences, engineering, science, and medicine and health sciences. Relevant research centers include CESAR, the OGRC, WRC, and the Center of Excellence in Marine Biotechnology (CEMB). These centers are in different stages of development: CESAR and WRC are fully active, the OGRC is working in a coordinating capacity, and CEMB is not fully staffed. Because these centers work across the university departments and with funding organizations, they often take a multidisciplinary approach, combining technical research with resource modeling, management, and policymaking.

The objectives of WRC are to synthesize research on water policy, science, and engineering with a focus on Omani and regional water issues.[43] Many of the center's projects are funded by MEDRC. Key research areas include those listed here:

- Socioeconomics and water resource management: Examining water markets, including pricing and subsidy schemes, increasing efficiency of cropping patterns, and irrigation and management of ground and surface water
- Desalination: Improving the efficiency of desalination, hybrid desalination plants, and use of renewable energy sources for desalination

---

[43] SQU, undated (b).

- Surface-water systems: Seawater-quality monitoring and management, water flow management, coastal-zone management, and effects that climate change has on surface water.

CESAR was established in 2000 to coordinate internally between university departments and externally with government ministries and private interests. Key areas of environmental research for this group are soil and water salinity, groundwater pollution, overgrazing, desertification, climate change and its consequences, and loss of habitat and biodiversity conservation.[44] In the GCC, CESAR has collaborated with the Gulf Research Council and the International Center for Biosaline Agriculture in Dubai. CESAR has also developed a biodiversity index for Oman, which could be useful for neighboring Gulf countries in identifying and monitoring their own biodiversity as part of their environmental protection efforts.

The OGCR currently coordinates research activities among the colleges and laboratories at SQU. The primary focus is applied research in oil-recovery techniques. Four key areas are being explored for new technologies and processes: thermal, chemical, microbial, and miscible recovery.[45]

The OGCR has a proposal for a new building, which it expects in time to be fully staffed and to have its own stand-alone labs. The OGCR is in discussion with Stanford Research Institute International in California to establish a joint oil-recovery center of excellence.[46]

The planned center of excellence in marine biotechnology is not yet operational, but it is also intended to play a coordinating role for departments of the university by bridging the gaps between basic and applied research and facilitating interdisciplinary efforts.[47] The prospective main areas of research focus for the center are micro- and macroalgae research for biofuels, bioprospecting of the marine envi-

---

[44] SQU, undated (a).

[45] Al-Maamari, 2010.

[46] Al-Maamari, 2010.

[47] SQU, undated (a).

ronment, and developing marine antifouling compounds.[48] CEMB hopes to build partnerships with other institutions in the region and globally for the advancement of marine biotechnology research.

Domestic human capital is limited; 75 percent of the staff at SQU are expatriates. Although recruiting faculty and researchers is challenging, retention is less of an issue. Much of the research funding comes from His Majesty's funds; Sultan Qaboos provides about OMR 0.5 million in grants annually; the university funds projects internally at about OMR 25,000 per project.[49] External funding typically comes from the petroleum sector. SQU has an agreement with UAEU for shared projects (50/50) on a rotational basis (i.e., one year, SQU submits the proposal, and UAEU submits the proposal the next year). Other regional collaborations include placing students at KISR in Kuwait. The Japan Cooperation Center, Petroleum (JCCP) also provides research grants. The university contributes in-kind by offering office space, transportation, and other assistance.

Middle East Desalination Research Center

MEDRC, located in Muscat, Oman, was created in 1996 as part of the Middle East peace process.[50] Its primary mission is to "contribute to the achievement of peace and stability in the Middle East and North Africa by promoting and supporting the use of desalination."[51] The vision of the institution is not only to encourage the development of new technologies but also to make those technologies affordable and accessible, promote capacity building, and forge regional partnerships through communication and cooperation. MEDRC's advocacy role is not the only aspect that sets it apart in the mix of GCC institutions surveyed. Although facilities are in place in Oman for MEDRC staff to conduct applied research, MEDRC is currently only a funding organization for research.

---

[48] Soussi, 2010.

[49] Al-Maamari, 2010.

[50] Interview with deputy director of MEDRC, 2010.

[51] MEDRC, undated (b).

The institute currently seeks to fund applied research projects that offer cost-effective solutions and help to build technical capacity. It receives support from the government, industry, and private donors and is able to fund approximately $2 million annually in research grants. Typical award grants are in the range of $30,000–$500,000. Researchers submitting grant proposals must collaborate with a foreign partner in the MENA region. The researcher has full patent rights for any commercial product generated by the research (five patents have been issued from MEDRC-funded research), but MEDRC retains the right to publish the reports on its website.

MEDRC's clear focus and area of strength is in desalination and water-reuse technology. Of all of the projects it has funded since its inception, 57 percent relate to increased membrane efficiency and 16 percent to renewable energy, and 27 percent are in other research areas related to water.[52] The ten areas of research that MEDRC seeks to fund are as follows:[53]

- Thermal desalination: Simplified design and performance improvements
- Membrane desalination: Membrane module and process design, energy recovery in RO processes, pretreatment methods, scaling and fouling fundamentals, and process and ancillary equipment design
- Alternative desalination: Development and feasibility studies of new concepts for nontraditional desalination processes and feasibility studies of desalination concepts
- Operation and maintenance of conventional desalination plants: Improving operation, efficiency, and reliability
- Intakes and outfalls: Procedures for selecting appropriate intake and outfall systems based on site conditions and development of new intake and outfall systems
- Energy issues: Reduction of energy consumption and affordable alternative energy solutions for desalination

---

[52] MEDRC data.

[53] MEDRC, undated (c).

- Environment issues: Assessing the environmental impact of desalination plant effluents
- Hybridized systems: Developing hybrid processes and reducing capital, operation, and maintenance costs
- Certification programs: Promoting and developing codes of practice for desalination
- Assessment studies: Critical assessments of the state of desalination research and areas for opportunities and cost-saving.

MEDRC was established in part to improve collaboration and has been successful in developing some partnerships across the MENA region, particularly with funded research in Israel, Jordan, and Oman. The new headquarters building in Oman near SQU Water Research Centre is intended to strengthen further cooperation between MEDRC and the university.[54] However, most of MEDRC's funding to date has gone to partners outside the region (43 percent to EU institutions, 18 percent to North American, and 7 percent to Australian).[55] Finding regional partners and proposals is a challenge. MEDRC has not yet fully staffed its permanent headquarters and research facility.

Bahrain

As a small island nation with a desert climate and petroleum resources, Bahrain's research priorities are in the areas of energy and the environment. Until recently, Bahrain had not had a dedicated institution for applied energy or environmental research, although universities and think tanks have contributed to both basic and applied research. Recently, the Bahrain Centre for Strategic, International and Energy Studies (BCSIES) was established, but its research focus remains unclear. We visited BCSR, which has been a key player in Bahrain's energy and environment research in collaboration with Arabian Gulf University and with other international institutions. BCSR is now

---

[54] MEDRC, undated (a).

[55] MEDRC, undated (a).

dissolved; however, we chose to profile the institution here because it provides an alternative think tank model for energy and environment research.

Bahrain Center for Studies and Research

BCSR was established in 1981 as an independent think tank for scientific applied research and studies. A program for strategic studies was added in 2003. Activities of the center included contractual research, training and capacity-building courses, symposia and seminars, and a biennial award for scientific research. BCSR also published a quarterly strategic journal. The center was dissolved under a royal decree in the summer of 2010.[56]

The organizational structure of BCSR consisted of three main strategic units: the Economic and Strategic Studies Unit; the Marketing, Social and Educational Studies Unit; and the Scientific Studies Unit. Each unit was headed by an assistant secretary-general who reported directly to the secretary-general; each was responsible for multiple research sections. The Scientific Studies Unit had three sections: Nutrition Studies, Environmental and Industrial Studies, and Fisheries Studies.

As a think tank, BCSR's research outputs typically fell on resource management and public policy. The bulk of its work was in economics and social studies. Key scientific research areas were the following four:

- marine life
- food security, nutrition, and health
- environmental pollution
- water resources.

Most of the center's funding (about 70 percent) came from the government of Bahrain, the remainder from private donations and contracted research with public and private organizations. Clients for research in energy and the environment included Bahrain's Public

---

[56] The center was dissolved shortly after the RQPI team's visit. The exact reason is unknown, but it is possible that the newly established BCSIES will take its place (Hanratty, 2010).

Commission for Protection of Marine Resources, Environment and Wildlife and the Electricity and Water Authority.

BCSR cooperated locally with Arabian Gulf University, Bahrain's Geomatec Spatial Information Research Center, and regionally with the Gulf Research Center in Dubai. Internationally, BCSR cooperated with organizations in Japan and China. It conducted an annual Japan–Islam symposium and had a memorandum of understanding with China in the China–Arab Friendship Association. BCSR was a member of Gallup Polls International. It also had a joint program with Lancaster University for capacity building.

At the time of the site visit, BCSR was assisting the European Commission in a three-year science and technology project to build a GCC network for science and technology research. BCSR was conducting its own research survey to assist this effort.

The Kingdom of Saudi Arabia

KSA has been involved in energy research since the discovery of massive oil reserves in the early 1960s. Until recently, most of its research in energy and the environment has been focused on applications for exploration, extraction, and operations within the petroleum industry. In the past two decades, Saudi Arabia's research priorities have expanded to include technologies for sustainable development and environmental management. Saudi Arabia has a large number of academic institutions and research centers compared with other Gulf countries. We chose to profile the three most-prominent institutions in energy and environment research within the country: KACST, KAUST, and KFUPM.[57]

King Abdulaziz City for Science and Technology
KACST has been in existence since 1977 and functions as the country's national science agency, houses the national laboratories, and

---

[57] We were unable to arrange for a visit to any institutions in the KSA, so all information in these profiles was collected from publicly available information.

manages a national science fund through annual grant programs.[58] KACST is responsible for applied research, technology development, and technology transfer, as well as the development of strategic plans and national policies for science and technology. KACST has outlined the following 14 strategic areas for the development of new and applied technologies:[59]

- water
- oil and gas
- petrochemicals
- nanotechnologies
- biotechnology
- IT
- electronics, communication, and photonics
- space and aeronautics
- energy
- advanced materials and the environment
- mathematics and physics
- medical and health
- agriculture technology
- building and construction.

Internally, KACST seeks to forge links between academic institutions, government agencies, and industries to take advantage of synergies. It supports the development of technology clusters and business and innovation incubators. KACST is a member of a variety of Arab, Islamic, and international associations.

King Abdullah University of Science and Technology

KAUST was officially inaugurated in 2009 as a graduate research university. Sustainability is at the core of KAUST's overall mission and research agenda. This focus extends to its architecture; the buildings use solar thermal and PV energy to produce up to 3,300 MWh of

---

[58] KACST, undated.

[59] KACST, undated.

electricity annually. Courtyards are designed to promote natural venti-lation.[60] Other considerations in the design of the campus were water conservation, green and sustainable building materials, energy effi-ciency and use of renewables, lighting and ventilation, and protection of nearby coral reefs and mangroves.

KAUST operates nine research centers, which are the primary research units for the university. These nine centers are multidisci-plinary in nature and focus on the following topic areas with applica-tions for energy and environment:[61]

- Catalysis: Novel use of CO_2 by catalysis, photocatalysis for hydro-gen generation from water, and solar-induced chemical reactions
- Computational bioscience: Search for bioactive molecules from Red Sea species
- Geometric modeling and scientific visualization: Support of mathematical modeling and computer imaging for other centers
- Advanced membranes and porous materials: Advanced separation technologies for water treatment and gas separation, liquid super-absorbents, and membrane optimization
- Plant-stress genomics: Improving crop yields under high salinity and frequent drought conditions
- Solar and Photovoltaics Engineering Research Center: Low-cost, disruptive PV technologies and innovative processing techniques
- Red Sea: Cataloging Red Sea biodiversity, distribution of pollut-ants, coral-reef ecology, and sustainability
- Clean combustion: Alternative fuels, CO_2 capture, gasification technologies, emission monitoring, and reduction technologies
- Water desalination and reuse: Hybrid membrane desalination and membrane technologies for potable reuse.

KAUST has a specialized center for marine and oceanic research in partnership with the Woods Hole Oceanographic Institution in the United States. It partners with IBM to host a supercomputing research

---

[60] KAUST, undated (b).

[61] KAUST, undated (a).

center with one of the most-advanced supercomputers in an academic environment globally. KAUST also partners with industry leaders, such as Saudi Arabian Oil Company (Saudi Aramco) and the Saudi Basic Industries Corporation (SABIC), and academic institutions in the United States, Europe, and East Asia. Regionally, it is in discussions with MEDRC for possible collaboration.

King Fahd University of Petroleum and Minerals

KFUPM was established to support the country's petroleum and mineral industries in Dhahran in 1963, when Saudi Arabia's first major oil field was discovered at Dammam Well 7.[62] It was first established as a college; it became a university in 1975. KFUPM offers both undergraduate and graduate degrees and carries out both basic and applied research. Applied research in the energy and environment arena is conducted in the following seven research centers:

- Center of Research Excellence in Corrosion
- Center of Research Excellence in Renewable Energy
- Center of Research Excellence in Petroleum Refining and Petrochemicals
- Center of Research Excellence in Nanotechnology
- Center for Environment and Water
- Center for Petroleum and Minerals
- Center for Engineering Research.

The needs of the country guide the research priorities.[63] Although, historically, the primary focus has been research in support of the petroleum industry, current research has expanded to include sustainable and renewable energy technologies. Specific research focus areas include the following:

---

[62] KFUPM, undated.

[63] KFUPM, undated.

- Water: Water resource development, planning, management and conservation, irrigation water management, water resource system modeling
- Solar: Nanotechnology applications in solar cells
- Environment: Environmental monitoring and pollution control, environmental modeling, waste treatment and management, environmental chemistry, baseline biological and oceanographic data collection, monitoring marine habitats in the Arabian Gulf and the Red Sea, and assessing human-made and natural impacts on local marine environments
- Energy: Nanotechnology and catalysis, CO_2 conversion to methanol, fuel cell development, economics of renewable energy, advanced energy storage, electrical infrastructure and control systems, and oil and gas engineering technologies.

KFUPM researchers collaborate with other institutions in the country, including KACST and Saudi Aramco.

Interview Protocol for the Gulf Cooperation Council Survey

This appendix contains the interview protocol for our survey.

Institutional Information

What is the mission/purpose for your institute?

1. What are the main activities of your institute (e.g., basic research, applied research, tech transfer)?
2. What are your key research areas in renewable and nonrenewable energy and environment?
    a. What type of facilities do you have for this research?
    b. How many research staff members do you have working at your institute?
3. Do you have any partner institutions (locally, regionally, or worldwide)?

Project Information

1. What major energy and environment projects are you currently undertaking or have planned in the near term?
    a. How are these projects funded?
    b. What are the expected outputs from these projects (e.g., scientific papers, patents, commercialization)?

2. Do you fund or collaborate on any projects outside of the GCC?

Additional Information and Recommendations

1. What are some of the challenges you have faced as an institution (e.g., funding, hiring researchers, facilities)?
   a. Any lessons learned?
2. What do you see as the most-promising areas of environment and energy research for the GCC?
   a. Are there any gaps that need to be filled?
   b. Are there opportunities for strategic partnerships? Where?
3. Is there anyone else with whom we should talk?
4. Are there areas that we did not cover that we should have?

Literature-Review Search Terms

We identified 60 total publications in our literature review that were relevant to energy and environment research in the Gulf region, as categorized in Table E.1.[1]

In our literature review, we used online databases and search terms listed in Table E.2.

Table E.1
Source Types for Literature Review

Literature Category	Number of Sources
University-based work in scholarly journals	33
Work by nonuniversity research organizations	11
Books	7
News media	12
Databases	4

---

[1] Some publications fell into more than one category.

Table E.2
Database Searches

Database	Search Terms
GreenFILE (EBSCO)	Gulf Cooperation Council OR Middle East AND renewable energy OR energy OR pollution OR wind power OR environment
	Pollution OR energy OR environment AND Middle East OR Levant OR North Africa
	Saudi Arabia OR Morocco OR Iran AND solar OR thermal OR wind AND energy
	Bahrain OR Israel OR Kuwait AND desalination OR desertification OR water
Earth and environmental science ebooks	Gulf Cooperation Council + subject: environment
	Middle East and subject: earth and environmental science
	Energy and subject: earth and environmental science
ISI Web of Knowledge	Gulf states AND energy OR green OR environment Subject areas selected: water resources, environmental engineering, multidisciplinary geosciences, energy and fuels, environmental studies, multidisciplinary agriculture, physical geography
	UAE OR Abu Dhabi OR Dubai AND renewable energy and environment Subject area selected: environmental sciences

References

Aboulnaga, Mohsen M., "A Roof Solar Chimney Assisted by Cooling Cavity for Natural Ventilation in Buildings in Hot Arid Climates: An Energy Conservation Approach in Al-Ain City," *Renewable Energy*, Vol. 14, No. 1–4, May–August 1998, pp. 357–363.

————, "Towards Green Buildings: Glass as a Building Element—The Use and Misuse in the Gulf Region," *Renewable Energy*, Vol. 31, No. 5, April 2006, pp. 631–653.

Adam, Jennifer C., and Dennis P. Lettenmaier, "Adjustment of Global Gridded Precipitation for Systematic Bias," *Journal of Geophysical Research*, Vol. 108, 2003, p. 4257.

Adhikari, Sushil, and Sandun Fernando, "Hydrogen Membrane Separation Techniques," *Industrial and Engineering Chemistry Research*, Vol. 45, No. 3, 2006, pp. 875–881.

Al Kuwari, Omran, chief executive officer, GreenGulf, interview with the authors, May 17, 2010.

Ala-Juusela, Mia, ed., *Heating and Cooling with Focus on Increased Energy Efficiency and Improved Comfort: Guidebook to IEA ECBCS Annex 37 Low Exergy Systems for Heating and Cooling of Buildings*, Espoo: VTT, 2004. As of January 28, 2011: http://lowex.org/english/inside/guidebook.html

Al-Derham, Hassan Rashid, vice president for research, Office of Academic Research, Qatar University, interview with authors, June 6, 2010.

Ali, Hikmat H., and Saba F. Al Nsairat, "Developing a Green Building Assessment Tool for Developing Countries: Case of Jordan," *Building and Environment*, Vol. 44, No. 5, May 2009, pp. 1053–1064.

Al-Jayyousi, Odeh R., "Greywater Reuse: Towards Sustainable Water Management," *Desalination*, Vol. 156, No. 1–3, August 1, 2003, pp. 181–192.

Allison, G. B., G. W. Gee, and S. W. Tyler, "Vadose-Zone Techniques for Estimating Groundwater Recharge in Arid and Semiarid Regions," *Soil Science Society of America Journal*, Vol. 58, 1994, pp. 6–14.

Al-Maamari, Rashid S., director, Oil and Gas Research Center, Sultan Qaboos University, interview with the authors, June 28, 2010.

Almehaideb, Reyadh, United Arab Emirates Faculty of Engineering, interview with the authors, May 11, 2010.

Al-Mohannadi, Hassan I., Chris O. Hunt, and Adrian P. Wood, "Controlling Residential Water Demand in Qatar: An Assessment," *Ambio*, Vol. 32, No. 5, August 2003, pp. 362–366.

Al-Mulla, Ali H., Azhari F. M. Ahmed, and Diane Lecoeur, *Qatar Photochemical Modelling Platform: A New Tool to Optimize Air Pollution Control for the Oil and Gas Industries*, Doha: International Petroleum Technology Conference, December 7–9, 2009.

Al-Mutairi, Naji, director general, Kuwait Institute for Scientific Research, interview with Obaid Younossi and Kristy Kamarck, June 3, 2010.

Al-Mutaz, Ibrahim S., "Hybrid RO MSF Desalination: Present Status and Future Perspectives," paper presented at the International Forum on Water, Resources, Technologies and Management in the Arab World, University of Sharjah, Sharjah, United Arab Emirates, May 8–10, 2005.

Appleyard, David, "Integrating Solar: CSP and Gas Turbine Hybrids," *Renewable Energy World*, May 28, 2010a. As of January 26, 2011:
http://www.renewableenergyworld.com/rea/news/article/2010/05/integrating-solar-gas-turbines

———, "Chilling Out in the Sun: Solar Cooling," *Renewable Energy World*, June 7, 2010b. As of January 26, 2011:
http://www.renewableenergyworld.com/rea/news/article/2010/06/chilling-out-in-the-sun-solar-cooling

Aquaterra Environmental Solutions, *Environmental Assessment and Action Plan for Qatari Environment (Draft)*, Ottawa, Ont., 2002.

Ashcroft, A. T., A. K. Cheetham, M. L. H. Green, and P. D. F. Vernon, "Partial Oxidation of Methane to Synthesis Gas Using Carbon Dioxide," *Nature*, Vol. 352, July 18, 1991, pp. 225–226.

Ashton, Peter J., Anthony R. Turton, and Dirk J. Roux, "Exploring the Government, Society, and Science Interfaces in Integrated Water Resource Management in South Africa," *Journal of Contemporary Water Research and Education*, Vol. 135, No. 1, December 2006, pp. 28–36.

Aziz, Ozair, Arif Inam, and Samiullah, "Utilization of Petrochemical Industry Waste Water for Agriculture," *Water, Air, and Soil Pollution*, Vol. 115, No. 1–4, October 1999, pp. 321–335.

Aziz, Ozair, M. Manzar, and Arif Inam, "Suitability of Petrochemical Industry Wastewater for Irrigation," *Journal of Environmental Science and Health*, Vol. 30, No. 4, 1995, pp. 735–751.

Banks, P. A., "Wastewater Reuse Case Studies in the Middle East," *Water Science Technology*, Vol. 23, 1991, pp. 2141–2148.

Bartis, James T., *Long-Range Energy R&D: A Methodology for Program Development and Evaluation*, Santa Monica, Calif.: RAND Corporation, TR-112-NETL, 2004. As of January 28, 2011:
http://www.rand.org/pubs/technical_reports/TR112.html

Beckman, James R., *Dewvaporation Desalination 5,000-Gallon-Per-Day Pilot Plant*, Denver, Colo.: U.S. Department of the Interior, Bureau of Reclamation, Technical Service Center, Water and Environmental Services Division, Water Treatment Engineering Research Team, Desalination and Water Purification Research and Development Program Report 120, June 2008. As of July 19, 2011:
http://www.usbr.gov/pmts/water/publications/reportpdfs/report120.pdf

Bell, Malcolm, and Robert Lowe, "Energy Efficient Modernisation of Housing: A UK Case Study," *Energy and Buildings*, Vol. 32, No. 3, September 2000, pp. 267–280.

Ben-Haim, Yakov, *Info-Gap Decision Theory: Decisions Under Severe Uncertainty*, Oxford: Academic, 2006.

Bin, Shui, and Hadi Dowlatabadi, "Consumer Lifestyle Approach to US Energy Use and the Related CO_2 Emissions," *Energy Policy*, Vol. 33, No. 2, January 2005, pp. 197–208.

Bishop, J. M., Y. Ye, A. H. Alsaffar, H. M. Al-Foudari, and S. Al-Jazzaf, "Diurnal and Nocturnal Catchability of Kuwait's Commercial Shrimps," *Fisheries Research*, Vol. 94, No. 1, October 2008, pp. 58–72.

British Petroleum, *BP Statistical Review of World Energy*, London, June 2010. As of December 15, 2010:
http://www.bp.com/liveassets/bp_internet/globalbp/globalbp_uk_english/reports_and_publications/statistical_energy_review_2008/STAGING/local_assets/2010_downloads/statistical_review_of_world_energy_full_report_2010.pdf

California Department of Water Resources, *Proposition 84 and Proposition 1e: Integrated Regional Water Management Guidelines*, 2010.

Cardwell, Hal E., Richard A. Cole, Lauren A. Cartwright, and Lynn A. Martin, "Integrated Water Resources Management: Definitions and Conceptual Musings," *Journal of Contemporary Water Research and Education*, Vol. 135, No. 1, December 2006, pp. 8–18.

Carnegie Mellon University, "EIO-LCA: Free, Fast, Easy Life Cycle Assessment," undated. As of December 17, 2010:
http://www.eiolca.net/

Carras, John N., Pamela M. Franklin, Yuhong Hu, A. K. Singh, Oleg V. Tailakov, David Picard, Azhari F. M. Ahmed, Eilev Gjerald, Susann Nordrum, and Irina Yesserkepova, "Fugitive Emissions," in Simon Eggleston, Leandro Buendia, Kyoko Miwa, Todd Ngara, and Kiyoto Tanabe, eds., *2006 IPCC Guidelines for National Greenhouse Gas Inventories*, Vol. 2: Energy, Hayama, Japan: Institute for Global Environmental Strategies, Intergovernmental Panel on Climate Change, 2006, pp. 4.1–4.78. As of January 28, 2011:
http://www.ipcc-nggip.iges.or.jp/public/2006gl/vol2.html

"CGC and BPL Global Win Kahramaa Project in Qatar," *T&D World*, January 21, 2009. As of January 28, 2011:
http://tdworld.com/info_systems/highlights/cgc-bpl-global-qatar-0109/

Chadha, Mridul, "World's Largest Concentrating Solar Power Plant to Come Up in Abu Dhabi," *Clean Technica*, June 12, 2010. As of January 31, 2011:
http://cleantechnica.com/2010/06/12/
worlds-largest-concentrating-solar-power-plant-to-come-up-in-abu-dhabi/

Chen, J. Paul, Lawrence K. Wang, and Lei Yang, "Thermal Distillation and Electrodialysis Technologies for Desalination," in Lawrence K. Wang, Yung-Tse Hung, and Nazih K. Shammas, eds., *Advanced Physiochemical Treatment Technologies*, Vol. 5, Totowa, N.J.: Humana Press, 2007, pp. 295–327.

CMU—*See* Carnegie Mellon University.

Crane, Keith, Aimee E. Curtright, David S. Ortiz, Constantine Samaras, and Nicholas Burger, "The Economic Costs of Reducing Greenhouse Gas Emissions Under a U.S. National Renewable Electricity Mandate," *Energy Policy*, Vol. 39, No. 5, May 2011, pp. 2730–2739.

Creyts, Jon, Anton Derkach, Scott Nyquist, Ken Ostrowski, and Jack Stephenson, *Reducing U.S. Greenhouse Gas Emissions: How Much at What Cost? U.S. Greenhouse Gas Abatement Mapping Initiative Executive Report*, New York: McKinsey and Company, December 2007. As of January 28, 2011:
http://www.mckinsey.com/clientservice/sustainability/greenhousegas.asp

Da Graça, G. Carrilho, Q. Chen, L. R. Glicksman, and L. K. Norford, "Simulation of Wind-Driven Ventilative Cooling Systems for an Apartment Building in Beijing and Shanghai," *Energy and Buildings*, Vol. 34, No. 1, January 2002, pp. 1–11.

Dalhuisen, Jasper M., Raymond J. G. M. Florax, Henri L. F. de Groot, and Peter Nijkamp, "Price and Income Elasticities of Residential Water Demand: A Meta-Analysis," *Land Economics*, Vol. 79, No. 2, May 2003, pp. 292–308.

Department of Agriculture and Water Research, Groundwater Unit, *Groundwater Data and Balance*, 2006a.

————, Irrigation and Drainage Unit, *Wells Water Survey in the Qatar Farms,* 2006b.

Dorica, J., P. Ramamurthy, and A. Elliott, *Reuse of Biologically Treated Effluents in Pulp and Paper Operations,* Pulp and Paper Research Institute of Canada, Report MR 372, 1998.

DWR—*See* California Department of Water Resources.

Eichholtz, Piet, Nils Kok, and John M. Quigley, *Why Do Companies Rent Green? Ecological Responsiveness and Corporate Real Estate,* Berkeley, Calif.: University of California, Berkeley, Institute of Business and Economic Research, Fisher Center for Real Estate and Urban Economics, Program on Housing and Urban Policy, Working Paper W09-004, July 2010. As of July 19, 2011:
http://urbanpolicy.berkeley.edu/pdf/
Who_Rents_Green_Journal_Version_NK071310.pdf

Electric Power Research Institute, *Comparison of Alternate Cooling Technologies for California Power Plants: Economic, Environmental and Other Tradeoffs,* 500-02-079F, February 2002a. As of August 24, 2011:
http://www.energy.ca.gov/reports/2002-07-09_500-02-079F.PDF

————, *Water and Sustainability,* Vol. 3: *U.S. Water Consumption for Power Production—The Next Half Century,* March 2002b. As of November 1, 2010:
http://mydocs.epri.com/docs/public/00000000001006786.pdf

————, *Program on Technology Innovation: Power Generation and Water Sustainability,* September 2007. As of November 1, 2010:
http://mydocs.epri.com/docs/public/000000000001015444.pdf

————, *The Green Grid: Energy Savings and Carbon Emissions Reductions Enabled by a Smart Grid, Technical Update,* June 2008.

El-Sorbagy, Abdel-moniem, "Design with Nature: Windcatcher as a Paradigm of Natural Ventilation Device in Buildings," *International Journal of Civil and Environmental Engineering,* Vol. 10, No. 3, 2010, pp. 26–31. As of December 17, 2010:
http://www.ijens.org/105403-6868%20IJCEE-IJENS.pdf

EMEC—*See* European Marine Energy Centre.

Energy Information Administration, "Electricity," undated.

————, "Greenhouse Gases, Climate Change, and Energy," last modified April 2, 2004. As of July 20, 2011:
http://www.eia.gov/oiaf/1605/ggccebro/chapter1.html

————, Office of Energy Markets and End Use, *Annual Energy Review,* August 19, 2010. As of July 20, 2011:
http://www.eia.gov/totalenergy/data/annual/

————, "Qatar," Country Analysis Briefs, last updated January 12, 2011a. As of December 15, 2010:
http://www.eia.doe.gov/emeu/cabs/Qatar/pdf.pdf

————, "Natural Gas Prices," last updated June 29, 2011b; referenced December 28, 2010. As of January 28, 2011:
http://www.eia.gov/dnav/ng/ng_pri_sum_dcu_nus_m.htm

EnerNex Corporation and WindLogics Inc., *Characterization of the Wind Resource in the Upper Midwest, Wind Integration Study: Task 1*, Xcel Energy and the Minnesota Department of Commerce, September 10, 2004. As of July 19, 2011:
http://www.uwig.org/XcelMNDOCwindcharacterization.pdf

EPA—*See* U.S. Environmental Protection Agency.

EPA, National Risk Management Research Laboratory, and U.S. Agency for International Development—*See* U.S. Environmental Protection Agency, Office of Wastewater Management, Municipal Support Division; National Risk Management Research Laboratory, Technology Transfer and Support Division; and U.S. Agency for International Development.

Eppler, Jeffrey, *Tidal Resource Characterization from Acoustic Doppler Current Profilers*, Seattle, Wash.: University of Washington, master's thesis, 2010.

EPRI—*See* Electric Power Research Institute.

Espey, M., J. Espey, and W. D. Shaw, "Price Elasticity of Residential Demand for Water: A Meta-Analysis," *Water Resources Research*, Vol. 33, No. 6, 1997, pp. 1369–1374.

European Commission Directorate-General for Climate Action, "Emissions Trading System (EU ETS)," last updated November 15, 2010. As of December 17, 2010:
http://ec.europa.eu/clima/policies/ets/index_en.htm

European Marine Energy Centre, undated home page. As of January 28, 2011:
http://www.emec.org.uk/

European Renewable Energy Council, undated home page. As of January 28, 2011:
http://www.erec.org/

Ewing, Reid H., *Pedestrian- and Transit-Friendly Design: A Primer for Smart Growth*, Washington, D.C.: Smart Growth Network, 1999.

"Expert: Water Consumption in Qatar Very High," *Middle East North Africa Financial Network*, March 19, 2009. As of January 28, 2011:
http://www.menafn.com/qn_news_story_s.asp?StoryId=1093239500

FAO—*See* Food and Agriculture Organization of the United Nations.

Fatta, D., Z. Salem, M. Mountadar, O. Assobhei, and M. Loizidou, "Urban Wastewater Treatment and Reclamation for Agricultural Irrigation: The Situation in Morocco and Palestine," *Environmentalist*, Vol. 24, No. 4, December 2004, pp. 227–236.

Fazeli, M. S., F. Khosravan, M. Hossini, S. Sathyanarayan, and P. N. Satish, "Enrichment of Heavy Metals in Paddy Crops Irrigated by Paper Mill Effluents Near Nanjangud, Mysore District, Karnatake, India," *Environmental Geology*, Vol. 34, No. 4, June 1998, pp. 297–302.

Food and Agriculture Organization of the United Nations, "Information on Fisheries Management in the State of Qatar," October 2003. As of July 19, 2011: http://www.fao.org/fi/oldsite/FCP/en/QAT/body.htm

———, "Qatar," 2005.

———, "Qatar," 2008. As of July 19, 2011: http://www.fao.org/nr/water/aquastat/countries/qatar/index.stm

———, *Irrigation in the Middle East Region in Figures: AQUASTAT Survey— 2008*, Karen Frenken, ed., Food and Agriculture Organization of the United Nations, Land and Water Division, 2009. As of July 19, 2011: http://www.fao.org/docrep/012/i0936e/i0936e00.htm

Forster, Piers, Venkatachalam Ramaswamy, Paulo Artaxo, Terje Berntsen, Richard Betts, David W. Fahey, James Haywood, Judith Lean, David C. Lowe, Gunnar Myhre, John Nganga, Ronald Prinn, Graciela Raga, Michael Schulz, and Robert Van Dorland, "Changes in Atmospheric Constituents and in Radiative Forcing," in Susan Solomon, Dahe Qin, Martin Manning, Melinda Marquis, Kristen Averyt, Melinda M. B. Tignor, Henry LeRoy Miller Jr., and Zhenlin Chen, eds., *Climate Change 2007: The Physical Science Basis—Contribution of Working Group I to the Fourth Assessment Report of the Intergovernmental Panel on Climate Change*, Cambridge, UK: Cambridge University Press, 2007, pp. 130–234. As of July 19, 2011: http://www.ipcc.ch/pdf/assessment-report/ar4/wg1/ar4-wg1-chapter2.pdf

Gas Processing Center, Qatar University, "GPC Research Program," undated (a). As of July 19, 2011: https://gpc.qu.edu.qa/ResearchProgram.aspx

———, "QU Gas Processing Center: Where Scholars and Experts Pave New Ways for Gas Processing," undated (b). As of July 20, 2011: https://gpc.qu.edu.qa/

"GCC Power 2010 to Focus on Smart Grid and New Trends," Qatar News Agency, 2010.

General Secretariat for Development Planning, "Qatar National Development Strategy, 2011–2016," undated (a). As of August 22, 2011: http://www.gsdp.gov.qa/portal/page/portal/GSDP_Vision_Root/GSDP_EN/ What%20We%20Do/Qatar%20National%20Strategy

————, "What We Do: Qatar National Vision 2030," undated (b). As of January 28, 2011:
http://www.gsdp.gov.qa/portal/page/portal/GSDP_Vision_Root/GSDP_EN/What%20We%20Do/QNV_2030

————, *Qatar National Vision 2030: Advancing Sustainable Development—Qatar's Second Human Development Report*, July 2009. As of January 24, 2011:
http://hdr.undp.org/en/reports/nationalreports/arabstates/qatar/QHDR_EN_2009.pdf

Gesser, Hyman D., Norman R. Hunter, and Chandra B. Prakash, "The Direct Conversion of Methane to Methanol by Controlled Oxidation," *Chemical Reviews*, Vol. 85, No. 4, August 1985, pp. 235–244.

Glennon, Robert, "The Price of Water," *Journal of Land, Resources, and Environmental Law*, Vol. 24, 2004, pp. 337–342.

Global Water Partnership, *Integrated Water Resources Management*, Stockholm, TEC Background Paper 4, March 2000.

Global Water Sustainability Center, home page, last updated June 16, 2011. As of January 24, 2011:
http://www.globalwsc.com/

Greenfield, Victoria A., Debra Knopman, Eric Talley, Gabrielle Bloom, Edward Balkovich, D. J. Peterson, James T. Bartis, Stephen Rattien, Richard A. Rettig, Mark Y. D. Wang, Michael Mattock, Jihane Najjar, and Martin C. Libicki, *Design of the Qatar National Research Fund: An Overview of the Study Approach and Key Recommendations*, Santa Monica, Calif.: RAND Corporation, TR-209-QF, 2008. As of July 19, 2011:
http://www.rand.org/pubs/technical_reports/TR209.html

GreenGulf, home page, undated. As of January 24, 2011:
http://green-gulf.com/en/

Gridwise Alliance, *What Is Missing in Our Fundamental Knowledge of Smart Grid Implementation?* November 9, 2009.

Grol, Eric, "Technical Assessment of an Integrated Gasification Fuel Cell Combined Cycle with Carbon Capture," *Energy Procedia*, Vol. 1, No. 1, February 2009, pp. 4307–4313.

Gross, Michael, "Algal Biofuel Hopes," *Current Biology*, Vol. 18, No. 2, 2008, pp. R46–R47.

Groves, David G., David Yates, and Claudia Tebaldi, "Developing and Applying Uncertain Global Climate Change Projections for Regional Water Management Planning," *Water Resources Research*, Vol. 44, No. W12413, 2008.

GSDP—*See* General Secretariat for Development Planning.

Gude, Veera Gnaneswar, Nagamany Nirmalakhandan, and Shuguang Deng, "Renewable and Sustainable Approaches for Desalination," *Renewable and Sustainable Energy Reviews*, Vol. 14, No. 9, December 2010, pp. 2641–2654.

GWP—*See* Global Water Partnership.

GWSC—*See* Global Water Sustainability Center.

Hamed, Osman A., "Overview of Hybrid Desalination Systems: Current Status and Future Prospects," *Desalination*, Vol. 186, 2006, pp. 207–214.

Hamed, Osman Ahmed, Mohammad A. K. Al-Sofi, Monazir Imam, Ghulam M. Mustafa, Khalid Ba-Mardouf, Hamed Al-Washmi, A. Al-Olyani, Nasir Al-Ameri, and A. Al-Zahrani, *Thermodynamic Analysis of Al-Jubail Power/ Water Co-Generation Cycles*, Al-Jubail, Saudi Arabia: Saline Water Conversion Corporation, Technical Report TR 3808/APP 98002, November 2000. As of July 19, 2011:
http://www.swcc.gov.sa/files/assets/Research/Technical%20Papers/Thermal/
THERMODYNAMIC%20ANALYSIS%20OF%20%20AL-JUBAIL%20
POWER%20WATER%20CO-GENERATI.pdf

Hanemann, W. M., "The Economic Conception of Water," in Peter P. Rogers, Manuel Ramón Llamas, and Luis Martínez-Cortina, eds., *Water Crisis: Myth or Reality?* London: Taylor and Francis, 2006, pp. 61–91.

Hanratty, Tom, "Staff Fear for Jobs," *Gulf Daily News*, June 24, 2010. As of January 24, 2011:
http://www.gulf-daily-news.com/NewsDetails.aspx?storyid=280905

Hanson, Blaine R., Jirka Šimůnek, and Jan W. Hopmans, "Evaluation of Urea-Ammonium-Nitrate Fertigation with Drip Irrigation Using Numerical Modeling," *Agricultural Water Management*, Vol. 86, No. 1–2, November 16, 2006, pp. 102–113.

He, Sufang, Hongmiao Wu, Wanjin Yu, Liuye Mo, Hui Lou, and Xiaoming Zheng, "Combination of CO_2 Reforming and Partial Oxidation of Methane to Produce Syngas over Ni/SiO_2 and $Ni–Al_2O_3/SiO_2$ Catalysts with Different Precursors," *International Journal of Hydrogen Energy*, Vol. 34, No. 2, January 2009, pp. 839–843.

Healey, P., "Building Institutional Capacity Through Collaborative Approaches to Urban Planning," *Environment and Planning A*, Vol. 30, No. 9, 1998, pp. 1531–1546.

Hileman, James I., David S. Ortiz, James T. Bartis, Hsin Min Wong, Pearl E. Donohoo, Malcolm A. Weiss, and Ian A. Waitz, *Near-Term Feasibility of Alternative Jet Fuels*, Santa Monica, Calif.: RAND Corporation, TR-554-FAA, 2009. As of January 28, 2011:
http://www.rand.org/pubs/technical_reports/TR554.html

Hooper, Bruce P., Geoffrey T. McDonald, and Bruce Mitchell, "Facilitating Integrated Resource and Environmental Management: Australian and Canadian Perspectives," *Journal of Environmental Planning and Management*, Vol. 42, No. 5, 1999, pp. 747–766.

Hosaka, Tomoko A., "Japan Looking to Sell 'Smart' Cities to the World," Associated Press, October 8, 2010.

Huovila, Pekka, Mia Ala-Juusela, Luciana Meichert, and Stéphane Pouffary, *Buildings and Climate Change: Status, Challenges, and Opportunities*, Paris: United Nations Environment Programme, Sustainable Consumption and Production Branch, 2007. As of July 19, 2011:
http://www.unep.fr/shared/publications/pdf/DTIx0916xPA-BuildingsClimate.pdf

International Energy Agency, *Coming in from the Cold: Improving District Heating Policy in Transition Economies*, Paris: International Energy Agency: Organisation for Economic Co-operation and Development, 2004.

———, Statistics Division, *Energy Balances of Non-OECD Countries 2007*, July 27, 2007. As of January 28, 2011:
http://www.oecd-ilibrary.org/energy/
energy-balances-of-non-oecd-countries-2007_energy_bal_non-oecd-2007-en-fr

International Monetary Fund, *World Economic Outlook Database: October 2010 Edition*, October 2010. As of January 28, 2011:
http://www.imf.org/external/pubs/ft/weo/2010/02/weodata/index.aspx

Ipakchi, Ali, and Farrokh Albuyeh, "Grid of the Future," *IEEE Power and Energy Magazine*, Vol. 7, No. 2, March–April 2009, pp. 52–62.

Jacobson, Allan J., "Materials for Solid Oxide Fuel Cells," *Chemistry of Materials*, Vol. 22, No. 3, 2010, pp. 660–674.

Jalalzadeh-Azar, Ali, Genevieve Saur, and Anthony Lopez, Biogas Resources Characterization, presentation, National Renewable Energy Laboratory 2010 Hydrogen Program Annual Merit Review, NREL/PR-560-48057, June 8, 2010. As of July 19, 2011:
http://www.nrel.gov/docs/fy10osti/48057.pdf

Jimenez-Cisneros, B., "Wastewater Reuse to Increase Soil Productivity," *Water Science and Technology*, Vol. 32, No. 12, 1995, pp. 173–180.

Johnson, Ted, "Battling Seawater Intrusion in the Central and West Coast Basins," *Water Replenishment District of Southern California*, Technical Bulletin, Vol. 13, Fall 2007. As of August 23, 2011:
http://www.wrd.org/engineering/seawater-intrusion-los-angeles.php

Jonathan Rose Companies, *Location Efficiency and Housing Type: Boiling It Down to BTUs*, revised March 2011. As of April 29, 2011:
http://www.epa.gov/smartgrowth/pdf/location_efficiency_BTU.pdf

Jønch-Clausen, Torkil, "Integrated Water Resources Management (IWRM) and Water Efficiency Plans by 2005: Why, What and How?" *Global Water Partnership*, TEC Background Paper 10, January 2004.

KACST—*See* King Abdulaziz City for Science and Technology.

"Kahramaa, BARWA, and Qatari Diar Research Institute Sign MOU," *Middle East North Africa Financial Network*, March 17, 2010. As of December 13, 2010: http://www.menafn.com/qn_news_story_s.asp?StoryId=1093314497

"Kahramaa Drive to Save Energy," *Peninsula*, June 7, 2010. As of July 20, 2011: http://www.thepeninsulaqatar.com/qatar/ 12229-kahramaa-drive-to-save-energy.html

Kalogirou, Soteris A., "Seawater Desalination Using Renewable Energy Sources," *Progress in Energy and Combustion Science*, Vol. 31, No. 3, 2005, pp. 242–281.

Kats, Gregory, Leon Alevantis, Adam Berman, Evan Mills, and Jeff Perlman, *The Costs and Financial Benefits of Green Buildings: A Report to California's Sustainable Building Task Force*, Sacramento, Calif.: Sustainable Building Task Force, October 2003. As of January 28, 2011: http://www.calrecycle.ca.gov/greenbuilding/Design/CostBenefit/Report.pdf

KAUST—*See* King Abdullah University of Science and Technology.

Keoleian, G. A., and W. A. Garner, *Industrial Ecology Educational Resources Compendium*, Ann Arbor, Mich.: National Pollution Prevention Center, University of Michigan, Ann Arbor, April 1995.

KFUPM—*See* King Fahd University of Petroleum and Minerals.

Khan, Nuzrat Yar, "Multiple Stressors and Ecosystem-Based Management in the Gulf," *Aquatic Ecosystem Health and Management*, Vol. 10, No. 3, 2007, pp. 259–267.

Khawaji, Akili D., Ibrahim K. Kutubkhanah, and Jong-Mihn Wie, "Advances in Seawater Desalination Technologies," *Desalination*, Vol. 221, No. 1–3, March 1, 2008, pp. 47–69.

King Abdulaziz City for Science and Technology, "Information About KACST," undated. As of January 24, 2011: http://www.kacst.edu.sa/en/about/Pages/default.aspx

King Abdullah University of Science and Technology, "Introduction," undated (a). As of January 24, 2011: http://www.kaust.edu.sa/research/centers/intro.html

———, "Sustainable Site Planning," undated (b). As of January 24, 2011: http://www.kaust.edu.sa/about/sustainable/planning.html

King Fahd University of Petroleum and Minerals, "About KFUPM: Location," undated; referenced January 24, 2011. As of July 19, 2011:
http://www.kfupm.edu.sa/SitePages/en/
DetailPage.aspx?CUSTOMID=4&LinkID=Link26

KISR—*See* Kuwait Institute for Scientific Research.

Krautkraemer, Jeffrey A., "Nonrenewable Resource Scarcity," *Journal of Economic Literature*, Vol. 36, No. 4, December 1998, pp. 2065–2107.

Kuwait Institute for Scientific Research, home page, undated. As of January 24, 2011:
http://www.kisr.edu.kw/

Kuwait University, Office of the Vice President of Research, *Research Administration: Annual Report 2008–09*, Kuwait, December 2009. As of January 24, 2011:
http://www.ovpr.kuniv.edu/research/publications/ann09_en.pdf

LaTourrette, Tom, Thomas Light, Debra Knopman, and James T. Bartis, *Managing Spent Nuclear Fuel: Strategy Alternatives and Policy Implications*, Santa Monica, Calif.: RAND Corporation, MG-970-RC, 2010. As of August 24, 2011:
http://www.rand.org/pubs/monographs/MG970.html

Lattemann, Sabine, and Thomas Hopner, "Impacts of Seawater Desalination Plants on the Marine Environment of the Gulf," in Abdulaziz H. Abuzinada, Hans-Jörg Barth, and Friedhelm Krupp, eds., *Protecting the Gulf's Marine Ecosystems from Pollution*, Basel: Birkhauser, January 2008, pp. 191–205.

Lempert, Robert J., David G. Groves, Steven W. Popper, and Steven C. Bankes, "A General, Analytic Method for Generating Robust Strategies and Narrative Scenarios," *Management Science*, Vol. 52, No. 4, April 2006, pp. 514–528.

Liebman, Jeffrey B., and Richard J. Zeckhauser, *Schmeduling*, Cambridge, Mass.: Harvard University and National Bureau of Economic Research, working paper, October 2004. As of July 19, 2011:
http://www.hks.harvard.edu/jeffreyliebman/schmeduling.pdf

Lisfet, Reid, "A Metaphor, a Field, and a Journal," *Journal of Industrial Ecology*, Vol. 1, No. 1, January 1997, pp. 1–3.

Litos Strategic Communication, *What a Smart Grid Means to Our Nation's Future*, Washington, D.C.: U.S. Department of Energy, undated.

———, *The Smart Grid: An Introduction*, Washington, D.C.: U.S. Department of Energy, 2008.

Logan, Gregg, Stephanie Siejka, and Shyam Kannan, *The Market for Smart Growth*, Atlanta, Ga.: Robert Charles Lesser and Company, undated. As of April 29, 2011:
http://www.epa.gov/smartgrowth/pdf/logan.pdf

Mansour, Maya, "50% Rise in Buildings in Qatar: Third Phase of Census 2010 Launched," *Qatar Tribune*, Vol. 4, No. 1325, April 21, 2010, p. 1. As of January 27, 2011:
http://www.qsa.gov.qa/eng/News/media_news/2010/21-apr/Qatar%20Tribune%2021-4-2010.pdf

Mansur, Erin T., and Sheila M. Olmstead, *The Value of Scarce Water: Measuring the Inefficiency of Municipal Regulations*, Cambridge, Mass.: National Bureau of Economic Research, Working Paper 13513, October 2007. As of July 19, 2011:
http://papers.nber.org/papers/13513

Maréchal, J. C., B. Dewandel, S. Ahmed, L. Galeazzi, and F. K. Zaidi, "Combined Estimation of Specific Yield and Natural Recharge in a Semi-Arid Groundwater Basin with Irrigated Agriculture," *Journal of Hydrology*, Vol. 329, No. 1–2, September 30, 2006, pp. 281–293.

Masdar Institute, "About Us," undated (a). As of January 24, 2011:
http://www.masdar.ae/en/Menu/Index.aspx?MenuID=42&mnu=Pri

———, "Fast Facts," undated (b); referenced February 1, 2011. As of July 19, 2011:
http://www.masdar.ac.ae/FastFacts.html

Maurer, E. P., J. C. Adam, and A. W. Wood, "Climate Model Based Consensus on the Hydrologic Impacts of Climate Change to the Rio Lempa Basin of Central America," *Hydrology and Earth System Sciences*, Vol. 13, 2009, pp. 183–194.

Mayer, Peter W., William B. DeOreo, E. M. Opitz, J. C. Kiefer, W. Y. Davis, B. Dziegielewski, and J. O. Nelson, eds., *Residential End Uses of Water*, Denver, Colo.: American Water Works Association, 1999.

MEDRC—*See* Middle East Desalination Research Center.

Meehl, Gerald A., Curt Covey, Thomas Delworth, Mojib Latif, Bryant McAvaney, John F. B. Mitchell, Ronald J. Stouffer, and Karl E. Taylor, "The WCRP CMIP3 Multimodel Dataset: A New Era in Climate Change Research," *Bulletin of the American Meteorological Society*, Vol. 88, No. 9, September 2007, pp. 1383–1394.

Metz, Bert, Ogunlade Davidson, Heleen de Coninck, Manuela Loos, and Leo Meyer, eds., *IPCC Special Report on Carbon Dioxide Capture and Storage*, Cambridge, UK: Cambridge University Press, 2005. As of July 19, 2011:
http://www.ipcc.ch/pdf/special-reports/srccs/srccs_wholereport.pdf

Middle East Desalination Research Center, *The Chairman's 10 Year Review*, undated (a). As of January 24, 2011:
http://www.medrc.org/about/Chairman10yreview.pdf

———, mission statement, undated (b). As of August 2, 2011:
http://www.medrc.org/index.cfm?area=about

———, "Research Program and Policy," undated (c). As of January 24, 2011:
http://www.medrc.org/index.cfm?area=research

—————, *The Middle East Desalination Research Center Review, 1997–2010*, c. 2011. Not available to the public.

Middlestadt, Susan, Mona Grieser, Orlando Hernandez, Khulood Tubaishat, Julie Sanchack, Brian Southwell, and Reva Schwartz, "Turning Minds On and Faucets Off: Water Conservation Education in Jordanian Schools," *Journal of Environmental Education*, Vol. 32, No. 2, Winter 2001, pp. 37–45.

Millennium Ecosystem Assessment, *Consolidation and Trends Working Group, Ecosystems and Human Well-Being: Synthesis*, Washington, D.C.: Island Press, 2005.

Minh, Nguyen Q., "Solid Oxide Fuel Cell Technology: Features and Applications," *Solid State Ionics*, Vol. 174, No. 1–4, October 29, 2004, pp. 271–277.

Misra, R. N., and P. K. Behera, "The Effect of Paper Industry Effluent on Growth, Pigments, Carbohydrates and Proteins of Rice Seedlings," *Environmental Pollution*, Vol. 72, No. 2, 1991, pp. 159–167.

Mitchell, Bruce, "Integrated Water Resource Management, Institutional Arrangements, and Land-Use Planning," *Environment and Planning A*, Vol. 37, No. 8, 2005, pp. 1335–1352.

Nakicenovic, Nebojsa, and Rob Swart, eds., *Special Report on Emissions Scenarios: A Special Report of Working Group III of the Intergovernmental Panel on Climate Change*, Cambridge, UK, 2000. As of July 29, 2011:
http://bibpurl.oclc.org/web/10394

Nataraj, Shanthi, and W. Michael Hanemann, "Does Marginal Price Matter? A Regression Discontinuity Approach to Estimating Water Demand," *Journal of Environmental Economics and Management*, Vol. 61, No. 2, March 2011, pp. 198–212.

National Energy Technology Laboratory, *Cost and Performance Baseline for Fossil Energy Plants*, Vol. 1: *Bituminous Coal and Natural Gas to Electricity Final Report*, DOE/NETL-2007-1281, original issue date May 2007, revision 1, August 2007a. As of July 19, 2011:
http://www.netl.doe.gov/energy-analyses/pubs/BitBase_FinRep_2007.pdf

—————, *The Benefits of SOFC for Coal-Based Power Generation*, Washington, D.C., October 30, 2007b. As of December 17, 2010:
http://www.netl.doe.gov/technologies/coalpower/fuelcells/publications/Final%20Report_OMB_Benefits%20of%20Fuel%20Cells_Coal%20Plant.pdf

—————, *Cost and Performance of Retrofitting Existing NGCC Units for Carbon Capture*, DOE/NETL-401-080610, October 1, 2010. As of January 28, 2011:
http://www.netl.doe.gov/energy-analyses/pubs/NGCC_Retrofit_Analysis.pdf

National Renewable Energy Laboratory, "Building Integrated Solar Technologies," last updated December 16, 2009. As of January 28, 2011:
http://www.nrel.gov/buildings/building_solar.html

————, "Parabolic Trough Power Plant System Technology," last updated January 28, 2010a. As of January 26, 2011:
http://www.nrel.gov/csp/troughnet/power_plant_systems.html

————, "U.S. Parabolic Trough Power Plant Data," last updated January 29, 2010b. As of July 22, 2011:
http://www.nrel.gov/csp/troughnet/power_plant_data.html

————, *National Renewable Energy Laboratory Sustainability Report FY 2009*, Golden, Colo., NREL/MP-3000-47450, November 2010c. As of July 19, 2011:
http://www.nrel.gov/docs/fy11osti/47450.pdf

————, home page, last updated July 15, 2011. As of January 28, 2011:
http://www.nrel.gov/

National Research Council, Committee on the Use of Treated Municipal Wastewater Effluents and Sludge in the Production of Crops for Human Consumption, *Use of Reclaimed Water and Sludge in Food Crop Production*, Washington, D.C.: National Academy Press, 1996. As of July 19, 2011:
http://www.nap.edu/catalog/5175.html

————, Committee on Advancing Desalination Technology, *Desalination: A National Perspective*, Washington, D.C.: National Academies Press, 2008.

————, Panel on Electricity from Renewable Resources, *Electricity from Renewable Resources: Status, Prospects, and Impediments*, Washington, D.C.: National Academies Press, 2010. As of January 28, 2011:
http://www.nap.edu/catalog.php?record_id=12619

Natural Gas Supply Association, "Residential Uses," undated web page. As of January 27, 2011:
http://www.naturalgas.org/overview/uses_residential.asp

"New Law to Help Boost Qatar's Marine Resources," *Doha Press*, July 3, 2010. As of August 23, 2011:
http://www.dohapress.com/portal/index.php/
archive/57-all-local-news/3021-new-law-to-help-boost-qatars-marine-resources

NREL—*See* National Renewable Energy Laboratory.

OECD—*See* Organisation for Economic Co-Operation and Development.

Olmstead, Sheila M., W. Michael Hanemann, and Robert N. Stavins, "Water Demand Under Alternative Price Structures," *Journal of Environmental Economics and Management*, Vol. 54, No. 2, September 2007, pp. 181–198.

Olmstead, Sheila M., and Robert N. Stavins, "Comparing Price and Nonprice Approaches to Urban Water Conservation," *Water Resources Research*, Vol. 45, No. 4, April 2009, pp. W04301–W04311.

Olson, Syanne, "Florida Power and Light Opens Hybrid Concentrated Solar Thermal/Natural Gas Power Plant," *PV-Tech*, March 7, 2011. As of July 22, 2011: http://www.pv-tech.org/news/florida_power_light_opens_hybrid_concentrated_solar_thermal_natural_gas_pow

Orera, A., and P. R. Slater, "New Chemical Systems for Solid Oxide Fuel Cells," *Chemistry of Materials*, Vol. 22, No. 3, 2010, pp. 675–690.

Organisation for Economic Co-Operation and Development, *The Measurement of Scientific and Technological Activities: Proposed Standard Practice for Surveys on Research and Experimental Development—Frascati Manual 2002*, Paris, 2002.

Organization of the Petroleum Exporting Countries, *Annual Statistical Bulletin 2009*, Vienna, Austria, 2010. As of December 16, 2010: http://www.opec.org/opec_web/static_files_project/media/downloads/publications/ASB2009.pdf

Ortiz, David S., Aimee E. Curtright, Constantine Samaras, Aviva Litovitz, and Nicholas Burger, *Near-Term Opportunities for Integrating Biomass into the U.S. Electricity Supply: Technical Considerations*, Santa Monica, Calif.: RAND Corporation, TR-984-NETL, 2011. As of August 24, 2011: http://www.rand.org/pubs/technical_reports/TR984.html

Pahl-Wostl, Claudia, "Transitions Towards Adaptive Management of Water Facing Climate and Global Change," *Water Resources Management*, Vol. 21, No. 1, 2007, pp. 49–62.

Parfomak, Paul W., and Aaron M. Flynn, *Liquefied Natural Gas (LNG) Import Terminals: Siting, Safety and Regulation*, Washington, D.C.: Congressional Research Service RL32205, January 28, 2004. As of January 28, 2011: http://www.au.af.mil/au/awc/awcgate/crs/rl32205.pdf

Park, Hi-Chun, and Eunnyeong Heo, "The Direct and Indirect Household Energy Requirements in the Republic of Korea from 1980 to 2000: An Input–Output Analysis," *Energy Policy*, Vol. 35, No. 5, May 2007, pp. 2839–2851.

Parker, Clinton E., and Syed R. Qasim, "Industrial Waste Management," in James R. Pfafflin and Edward N. Ziegler, eds., *Encyclopedia of Environmental Science and Engineering*, 5th ed., New York: Taylor and Francis, 2006, pp. 526–537.

Pearce, Fred, "Qatar to Use Biofuels? What About the Country's Energy Consumption?" *Guardian*, January 14, 2010. As of December 17, 2010: http://www.guardian.co.uk/environment/2010/jan/14/qatar-biofuels-energy-consumption

Pepermans, G., J. Driesen, D. Haeseldonckx, R. Belmans, and W. D'Haeseleer, "Distributed Generated: Definition, Benefits and Issues," *Energy Policy*, Vol. 33, No. 6, April 2005, pp. 787–798.

Permanent Population Committee, *Sustainable Development Indicators in the State of Qatar*, Issue 2, January 2010. As of July 19, 2011:
http://www.gsdp.gov.qa/portal/page/portal/ppc/PPC_home/PPC_Publications/studies/%D9%85%D8%A4%D8%B4%D8%B1%D8%A7%D8%AA.pdf

Pervin, Tanjima, Ulf-G Gerdtham, and Carl Hampus Lyttkens, "Societal Costs of Air Pollution–Related Health Hazards: A Review of Methods and Results," *Cost Effectiveness and Resource Allocation*, Vol. 6, No. 19, 2008.

Pöyry Energy, "Ras Laffan Industrial City Common Seawater Cooling System," undated. As of May 3, 2011:
http://www.poyry.ch/linked/en/aboutus/Ras_Laffan_Industrial_City.pdf

Prathapar, S. A., A. Jamrah, M. Ahmed, S. Al Adawi, S. Al Sidairi, and A. Al Harassi, "Overcoming Constraints in Treated Greywater Reuse in Oman," *Desalination*, Vol. 186, No. 1–3, December 30, 2005, pp. 177–186.

"Preserving Water, a Precious Natural Resource," *Peninsula*, May 3, 2010. As of July 20, 2011:
http://www.thepeninsulaqatar.com/q/56-tofol-jassim-al-nasr/1704-preserving-water-a-precious-natural-resource.html

Public Works Authority, *Treated Sewage Effluent*, 2005.

Puig-Bargués, J., G. Arbat, M. Elbana, M. Duran-Ros, J. Barragán, F. Ramírez de Cartagena, and F. R. Lamm, "Effect of Flushing Frequency on Emitter Clogging in Microirrigation with Effluents," *Agricultural Water Management*, Vol. 97, No. 6, June 2010, pp. 883–891.

"Qatar: Waste Management to Go Hi-Tech," *Middle East North Africa Financial Network*, April 7, 2008. As of July 19, 2011:
http://www.menafn.com/qn_news_story_s.asp?StoryId=1093191882

Qatar Electricity and Water Company, *Desalination of Water*, 2007.

Qatar Foundation, undated home page. As of January 28, 2011:
http://www.qf.org.qa/output/page3.asp

"Qatar GDP Statistics Show Gas Revenue Surpassed Oil in 2009," *Zawya*, 2010.

Qatar National Food Security Programme, "Mission," undated web page. As of January 28, 2011:
http://www.qnfsp.gov.qa/about-us/mission

Qatar National Research Fund, home page, undated. As of January 24, 2011:
http://www.qnrf.org

————, "Second Cycle: Development of Novel Gas-to-Liquid Technology in Near-Critical and Supercritical Phase Media," modified October 19, 2010. As of July 21, 2011:
http://www.qnrf.org/awarded/nprp/2/index.php?ELEMENT_ID=804

Qatar News Agency, "Environment," undated. As of December 17, 2010:
http://www.qnaol.net/QNAEn/Main_Sectors/Environment/Pages/Default.aspx

Qatar Science and Technology Park, "Qatar Petroleum Will Open Research Centre at QSTP," press release, April 1, 2008. As of November 1, 2010:
http://www.qstp.org.qa/output/page2071.asp

Qatar Sustainable Water and Energy Utilization Initiative, undated home page. As of August 1, 2011:
http://qwe.qatar.tamu.edu/

Qatar University, Office of Vice President for Research, Environmental Studies Center, "Welcome Message," last modified July 14, 2011a. As of December 17, 2010:
http://www.qu.edu.qa/offices/research/esc/

————, Office of Vice President for Research, Materials Technology Unit, "Materials Research Group," last modified July 14, 2011b. As of January 24, 2011:
http://www.qu.edu.qa/offices/research/MTU/materials_research_group.php

————, Office of Vice President for Research, Office of Academic Research, "Statistics," last modified July 15, 2011c. As of August 23, 2011:
http://www.qu.edu.qa/offices/research/academic/statistics.php

Qiblawey, Hazim Mohameed, and Fawzi Banat, "Solar Thermal Desalination Technologies," *Desalination*, Vol. 220, No. 1–3, March 2008, pp. 633–644.

QNFSP—*See* Qatar National Food Security Programme.

QNRF—*See* Qatar National Research Fund.

QSTP—*See* Qatar Science and Technology Park.

Rajannan, G., and G. Oblisami, "Effect of Paper Factory Effluents on Soil and Crop Plants," *Indian Journal of Environmental Health*, Vol. 21, No. 2, 1979, pp. 120–130.

Rees, Judith A., *Urban Water and Sanitation Services: An IWRM Approach*, Global Water Partnership, Technical Committee Background Paper 11, June 2006.

Rees, William E., "Building More Sustainable Cities," *Scientific American*, Vol. 19, No. 18, March 12, 2009.

Renwick, Mary E., and Richard D. Green, "Do Residential Water Demand Side Management Policies Measure Up? An Analysis of Eight California Water Agencies," *Journal of Environmental Economics and Management*, Vol. 40, No. 1, 2000, pp. 37–55.

Richardson, T. G., "Reclaimed Water for Residential Toilet Flushing: Are We Ready?" *Water Reuse Conference Proceedings*, Denver, Colo., 1998.

Richer, Renee, *Conservation in Qatar: Impacts of Increasing Industrialization*, Doha: Center for International and Regional Studies, Georgetown University School of Foreign Service in Qatar, 2008.

Ries, Charles P., Joseph Jenkins, and Oliver Wise, *Improving the Energy Performance of Buildings: Learning from the European Union and Australia*, Santa Monica, Calif.: RAND Corporation, TR-728-RER/BOMA, 2009. As of January 28, 2011:
http://www.rand.org/pubs/technical_reports/TR728.html

Roberts, Trenton L., Scott A. White, Arthur W. Warrick, and Thomas L. Thompson, "Tape Depth and Germination Method Influence Patterns of Salt Accumulation with Subsurface Drip Irrigation," *Agricultural Water Management*, Vol. 95, No. 6, June 2008, pp. 669–677.

Sahai, R., S. Jabeen, and P. K. Saxena, "Effect of Distillery Effluent on Seed Germination, Seedling Growth and Pigment Content of Rice," *Indian Journal of Ecology*, Vol. 10, 1983, pp. 7–10.

Sahai, R., N. Shukla, S. Jabeen, and P. K. Saxena, "Pollution Effect of Distillery Waste on the Growth Behavior of *Phaseolus radiatus L.*," *Environmental Pollution Series A*, Ecological and Biological, Vol. 37, No. 3, 1985, pp. 245–253.

Sambidge, Andy, "Doha Traffic Congestion Is Main Concern: Survey," *Arabian Business*, March 20, 2009. As of April 29, 2011:
http://www.arabianbusiness.com/
doha-traffic-congestion-is-main-concern-survey-64863.html

———, "Qatar Eyes Water Desalination Plant Expansion," *Arabian Business*, July 11, 2010. As of January 2011:
http://www.arabianbusiness.com/
qatar-eyes-water-desalination-plant-expansion-305936.html

Scheffler, T. B., and A. J. Leao, "Fabrication of Polymer Film Heat Transfer Elements for Energy Efficient Multi-Effect Distillation," *Desalination*, Vol. 222, No. 1–3, March 1, 2008, pp. 696–710.

Shelef, Gedaliah, "Wastewater Reclamation and Water Resources Management," *Water Science and Technology*, Vol. 24, No. 9, 1991, pp. 251–265.

Shuval, Hillel I., Avner Adin, Badri Fattal, Eliyahu Rawitz, and Perez Yekutiel, *Wastewater Irrigation in Developing Countries: Health Effects and Technical Solutions*, Washington, D.C.: World Bank, Technical Paper 51, 1986.

Simmers, I., ed., *Estimation of Natural Groundwater Recharge*, Dordrecht: D. Reidel Pub. Co., North Atlantic Treaty Organization ASI Series C, *Mathematical and Physical Sciences*, Vol. 222, 1987.

Singh, Amanjeet, Matt Syal, Sue C. Grady, and Sinem Korkmaz, "Effects of Green Buildings on Employee Health and Productivity," *American Journal of Public Health*, Vol. 100, No. 9, September 2010, pp. 1665–1668.

Sistla, Phanikumar V. S., G. Venkatesan, Purnima Jalihal, and S. Kathiroli, "Low Temperature Thermal Desalination Plants," *Proceedings of the Eighth (2009) ISOPE Ocean Mining Symposium*, Chennai, India, September 20–24, 2009. As of July 20, 2011:
http://www.isope.org/publications/proceedings/ISOPE_OMS/OMS%202009/papers/M09-83Sistla.pdf

Smith, V. Kerry, *Estimating Economic Values for Nature: Methods for Non-Market Valuation*, Brookfield, Vt.: Edward Elgar, 1996.

Snyder, Jerry K., A. K. Deb, Daniel A. Okun, Robert M. Clark, Jerry Snyder, Arun K. Deb, Walter M. Grayman, Frank M. Grablutz, Sandra B. McCammon, Scott M. Tyler, and Dagan Savic, *Impacts of Fire Flow and Distribution System Water Quality, Design, and Operation*, Denver, Colo.: AWWA Research Foundation and American Water Works Association, 2002.

Soussi, Bassam, United Nations Educational, Scientific and Cultural Organization Chair in Marine Biotechnology, Sultan Qaboos University, and chair, Center of Excellence in Marine Biotechnology, "Survey Questions on CEMB," email to Kristy Kamarck, RAND Corporation, June 27, 2010.

SQU—*See* Sultan Qaboos University.

Stambouli, A. Boudghene, and E. Traversa, "Solid Oxide Fuel Cells (SOFCs): A Review of an Environmentally Clean and Efficient Source of Energy," *Renewable and Sustainable Energy Reviews*, Vol. 6, No. 5, October 2002, pp. 433–455.

Sultan Qaboos University, Center of Excellence in Marine Biotechnology, "Objectives and Strategies," undated (a). As of January 24, 2011:
http://www.squ.edu.om/tabid/5175/language/en-US/Default.aspx

———, Water Research Center, "Objectives," undated (b). As of January 24, 2011:
http://www.squ.edu.om/tabid/9899/language/en-US/Default.aspx

Supreme Council for the Environment and Natural Reserves, *The National Report on Sustainable Development*, Doha, 2002.

Texas A&M University at Qatar, *Research at Texas A&M University at Qatar: Adding Knowledge to Qatar and the World*, undated.

Tietz, F., "Solid Oxide Fuel Cells," in K. H. Jürgen Buschow, Robert W. Cahn, Merton C. Flemings, Bernard Ilschner, Edward J. Kramer, Subhash Mahajan, and Patrick Veyssière, eds., *Encyclopedia of Materials Science and Technology*, Amsterdam: Elsevier, 2007, pp. 1–8.

Toman, Michael, Aimee E. Curtright, David S. Ortiz, Joel Darmstadter, and Brian Shannon, *Unconventional Fossil-Based Fuels: Economic and Environmental Trade-Offs*, Santa Monica, Calif.: RAND Corporation, TR-580-NCEP, 2008. As of January 28, 2011:
http://www.rand.org/pubs/technical_reports/TR580.html

Toman, Michael, James Griffin, and Robert J. Lempert, *Impacts on U.S. Energy Expenditures and Greenhouse-Gas Emissions of Increasing Renewable-Energy Use*, Santa Monica, Calif.: RAND Corporation, TR-384-1-EFC, 2008. As of August 24, 2011:
http://www.rand.org/pubs/technical_reports/TR384-1.html

Toumi, Habib, "Qatar Moves Towards Protecting Marine Resources and Boosting Fast Depleting Fish Stock," *Habib Toumi*, July 3, 2010. As of December 17, 2010:
http://www.habibtoumi.com/2010/07/03/qatar-moves-towards-protecting-marine-resources-and-boosting-fast-depleting-fish-stock/

UAEU—*See* United Arab Emirates University.

UCS—*See* Union of Concerned Scientists.

UNCCD—*See* United Nations Convention to Combat Desertification.

UNDP—*See* United Nations Development Programme.

Union of Concerned Scientists, *How Geothermal Energy Works*, last revised December 16, 2009. As of December 17, 2010:
http://www.ucsusa.org/clean_energy/technology_and_impacts/energy_technologies/how-geothermal-energy-works.html

United Arab Emirates University, "About the UAEU," last updated April 26, 2011; referenced January 24, 2011. As of July 20, 2011:
http://www.uaeu.ac.ae/about/

United Arab Emirates University staff and researchers, interview with the authors, May 11, 2010.

United Nations, Department of Policy Coordination and Sustainable Development, Division for Sustainable Development, *Country Profile: Qatar—Implementation of Agenda 21: Review of Progress Made Since the United Nations Conference on Environment and Development*, New York, 1997.

United Nations Convention to Combat Desertification, "FAQ," updated June 27, 2011.

United Nations Development Programme, *Human Development Report 2007/2008: Fighting Climate Change—Human Solidarity in a Divided World*, New York: Palgrave Macmillan, 2007.

Ürge-Vorsatz, Diana, L. D. Danny Harvey, Sevastianos Mirasgedis, and Mark D. Levine, "Mitigating CO_2 Emissions from Energy Use in the World's Buildings," *Building Research and Information*, Vol. 35, No. 4, July 2007, pp. 379–398.

U.S. Department of Energy, Advanced Research Projects Agency–Energy, "Direct Solar Fuels," undated (a); referenced January 26, 2011. As of July 20, 2011:
http://arpa-e.energy.gov/ProgramsProjects/OtherProjects/DirectSolarFuels.aspx

————, Advanced Research Projects Agency–Energy, "Electrofuels," undated (b). As of January 26, 2011:
http://arpa-e.energy.gov/ProgramsProjects/Electrofuels.aspx

————, *Basic Research Needs for Solar Energy Utilization: Report of the Basic Energy Sciences Workshop on Solar Energy Utilization, April 18–21, 2005*, Washington, D.C., revised September 2005. As of September 6, 2011:
http://science.energy.gov/~/media/bes/pdf/reports/files/seu_rpt_print.pdf

————, Building Technologies Program, *Building America Best Practices Series*, 2008.

————, Office of Energy Efficiency and Renewable Energy, "Solar Resource Characterization and Data Gathering," last updated August 3, 2010. As of December 17, 2010:
http://www1.eere.energy.gov/solar/characterization.html

————, Office of Fossil Energy "Solid State Energy Conversion Alliance," last updated January 31, 2011. As of December 17, 2010:
http://www.fossil.energy.gov/programs/powersystems/fuelcells/fuelcells_seca.html

U.S. Environmental Protection Agency, Office of Water, *Cases in Water Conservation: How Efficiency Programs Help Water Utilities Save Water and Avoid Costs*, Washington, D.C., 2002. As of July 20, 2011:
http://purl.access.gpo.gov/GPO/LPS50790

————, Clean Air Markets Division, Office of Air and Radiation, *Acid Rain Program: 2004 Progress Report*, EPA-430-R-05-012, October 2005. As of August 23, 2011:
http://www.epa.gov/airmarkets/progress/docs/2004report.pdf

————, "Particulate Matter (PM-10)," last modified April 10, 2010a. As of December 17, 2010:
http://www.epa.gov/airtrends/aqtrnd95/pm10.html

————, Offices of Air and Radiation, Research and Development, and Radiation and Indoor Air, "Indoor Air Facts No. 4 (revised) Sick Building Syndrome," last updated September 30, 2010b. As of January 28, 2011:
http://www.epa.gov/iaq/pubs/sbs.html

————, "Distribution Systems Research," last updated January 3, 2011; referenced November 19, 2010c. As of July 20, 2011:
http://www.epa.gov/nrmrl/wswrd/dw/dsr.html

————, "An Introduction to Indoor Air Quality (IAQ)," last updated November 29, 2010d. As of December 17, 2010:
http://www.epa.gov/iaq/is-imprv.html

———, "Green Building Research," last updated December 22, 2010e. As of January 27, 2011:
http://www.epa.gov/greenbuilding/pubs/about.htm#5

———, Office of Sustainable Communities, "About Smart Growth," last updated June 17, 2011. As of April 29, 2011:
http://www.epa.gov/smartgrowth/about_sg.htm

U.S. Environmental Protection Agency, Office of Wastewater Management, Municipal Support Division; National Risk Management Research Laboratory, Technology Transfer and Support Division; and U.S. Agency for International Development, *Guidelines for Water Reuse*, Washington, D.C.: U.S. Environmental Protection Agency and U.S. Agency for International Development, EPA/625/R-04/108, August 2004. As of July 19, 2011:
http://purl.access.gpo.gov/GPO/LPS81859

U.S. Government Accountability Office, *Energy–Water Nexus: Improvements to Federal Water Use Data Would Increase Understanding of Trends in Power Plant Water Use*, Washington, D.C., GAO-10-23, October 2009. As of July 20, 2011:
http://purl.access.gpo.gov/GPO/LPS120826

Van der Keur, P., H. J. Henriksen, J. C. Refsgaard, M. Brugnach, Claudia Pahl-Wostl, A. R. P. J. Dewulf, and Hendrik Buiteveld, "Identification of Major Sources of Uncertainty in Current IWRM Practice: Illustrated for the Rhine Basin," *Water Resources Management*, Vol. 22, No. 11, 2008, pp. 1677–1708.

Vangtook, Prapapong, and Surapong Chirarattananon, "Application of Radiant Cooling as a Passive Cooling Option in Hot Humid Climate," *Building and Environment*, Vol. 42, No. 2, February 2007, pp. 543–556.

Western Governors' Association, *Clean and Diversified Energy Initiative: Solar Task Force Report*, January 2006. As of January 26, 2011:
http://www.trec-uk.org.uk/resources/Solar_task_force_report.pdf

Winkler, Wolfgang, and Hagen Lorenz, "The Design of Stationary and Mobile Solid Oxide Fuel Cell–Gas Turbine Systems," *Journal of Power Sources*, Vol. 105, No. 2, March 20, 2002, pp. 222–227.

Wood, A. W., L. R. Leung, V. Sridhar, and D. P. Lettenmaier, "Hydrologic Implications of Dynamical and Statistical Approaches to Downscaling Climate Model Outputs," *Climatic Change*, Vol. 62, No. 1–3, 2004, pp. 189–216.

World Bank, "Population, Total," undated. As of August 22, 2011:
http://data.worldbank.org/indicator/SP.POP.TOTL/countries/QA?display=graph

———, *World Development Indicators*, April 22, 2010. As of July 20, 2011:
http://data.worldbank.org/news/world-development-indicators-2010-released

World Resources Institute, "Energy and Resources: Searchable Database," EarthTrends: The Environmental Information Portal, undated database. As of January 28, 2011:
http://earthtrends.wri.org/searchable_db/index.php?theme=6

Zaidi, S. M. Javaid, S. U. Rahman, and Halim H. Zaidi, "R&D Activities of Fuel Cell Research at KFUPM," *Desalination*, Vol. 209, No. 1–3, 2007, pp. 319–327.

Zekri, Slim, "Using Economic Incentives and Regulations to Reduce Seawater Intrusion in the Batinah Coastal Area of Oman," *Agricultural Water Management*, Vol. 95, No. 3, March 2008, pp. 243–252.

Zhai, Haibo, Edward S. Rubin, and Peter L. Versteeg, "Water Use at Pulverized Coal Power Plants with Postcombustion Carbon Capture and Storage," *Environmental Science and Technology*, Vol. 45, No. 6, March 15, 2011, pp. 2479–2485.